Universitext

For other titles in this series, go to
www.springer.com/series/223

Marino Badiale · Enrico Serra

Semilinear Elliptic Equations for Beginners

Existence Results via the Variational Approach

 Springer

Marino Badiale
Dipartimento di Matematica
Università di Torino
Torino, Italy
marino.badiale@unito.it

Enrico Serra
Dipartimento di Matematica
Università di Torino
Torino, Italy
enrico.serra@polito.it

ISBN 978-0-85729-226-1   e-ISBN 978-0-85729-227-8
DOI 10.1007/978-0-85729-227-8
Springer London Dordrecht Heidelberg New York

British Library Cataloguing in Publication Data
A catalogue record for this book is available from the British Library

Mathematics Subject Classification (2000): 35-01, 35A15, 35B38, 35J15, 35J20, 35J25, 35J60, 35J61, 35J92

Cover design: deblik

Printed on acid-free paper

Springer is part of Springer Science+Business Media (www.springer.com)

Preface

This book is an *introduction* to the study of a wide class of partial differential equations, namely semilinear elliptic problems, through the application of variational methods and critical point theory. Semilinear elliptic equations arise in a variety of contexts in geometry, physics, mechanics, engineering and, more recently, in life sciences. None of these fields can be investigated without taking into account nonlinear phenomena, and variational methods, as a branch or an evolution of the Calculus of Variations, are almost entirely concerned with nonlinearity. The theory has attained a spectacular success in the last thirty-or-so years, reaching a high level of complexity and refinement, and its applications are scattered throughout thousands of research papers. Furthermore, some of the simplest methods within the variational approach are nowadays classical tools in the field of nonlinear differential equations as fixed point and monotonicity methods are. They have become, in other words, part of the toolbox that any researcher in the field is supposed to have at hand.

This poses a stimulating problem: how to *teach* variational methods and semilinear elliptic equations to upcoming generations of students. In our teaching experience, we have noticed a recurrent phenomenon. Up to a certain grade, students possess a rather general knowledge of basic principles, for instance in Functional Analysis. At the next step however they are often confronted with problems taken from real research. This creates a gap that students generally fill in by themselves, following a path of hard work that requires a great deal of commitment and self-discipline.

Both of us have been teaching courses on the topics collected in this book for a number of years in various Italian universities, and the didactic demand that we encountered was almost always located in the gap described above. We found that the many excellent existing books on variational methods and critical point theory (e.g. [2, 18, 26, 35, 43, 45, 48]), all of them deservedly well-known internationally, are sometimes problematic for classroom use, precisely because they are too complete, or too long, or too advanced.

Hence, the *purpose* of this book, derived directly from classroom experience, is that of providing a support to the students and the teacher engaged in a first course in semilinear elliptic equations. We have tried to write a *textbook* that matches what we normally write on the blackboard in class. This is why the topics are introduced gradually, and in certain cases some redundancy is purposely added, to stress the fundamental steps in the building of the theory. As a consequence, in this book the reader will not find any sophisticated results, or fancy applications, or fashionable examples, but rather a first overview on the subject, in a sort of "elliptic equations for the layman", where *all* the details are written down. Every abstract result is immediately put into action to show how it works to solve an elliptic problem, reflecting the fact that the theory was developed with the intent of treating ever larger classes of equations. We believe that in this way the reader will be able not only to have the tools at hand, but also to see how they really act when applied to concrete problems. In the applications, we limited ourselves to the discussion of Dirichlet boundary value problems. While this is certainly reductive, it provides nonetheless a unified thread throughout the book, and has the advantage of introducing the student to the most common type of boundary conditions. Other types of problems can then be understood with a minimal effort, and the existing literature provides extensions of all sorts for the interested reader.

The contents of this book require a general preparation in analysis, including some notions of Functional Analysis, and some knowledge of the most common function spaces, such as Lebesgue and Sobolev spaces. These are all notions usually taught in undergraduate courses throughout the world. In any case, whenever we quote a result without proving it, we provide precise references for consultation.

These notes are to be *used* by two possibly distinct categories of students: those who need to know how some types of nonlinear problems can be handled, and who need the basic working tools from the variational approach and critical point theory, as well as students who are oriented towards research in the field of nonlinear differential equations. Clearly if for the former this book may be the last they read on the subject, for the latter it should be just the first. We set the level of these notes between a graduate course and a first year PhD course. Mathematicians, physicists and engineers are the natural audience this book is addressed to.

The material is divided into four chapters. The first one is an introduction, and contains a review of differential calculus for functionals, with many examples, and a few basic facts from the linear theory that will be used throughout the book. Convexity arguments complete the discussion providing the first examples of existence theorems.

Chapter 2 introduces the fundamentals of minimization techniques, for it is well-known that the simplest way to obtain a critical point of a functional is to look for a global extremum, which in most of the cases is a global minimum. However, for indefinite (unbounded) functionals, although global minimum points cannot exist, minimization techniques can still be profitably used by constraining the functional on a set where it is bounded from below. Typical examples of such sets are spheres or the Nehari manifold. The unifying feature of this chapter is the fact that all problems share some compactness properties that simplify the convergence arguments. A final section contains some examples of quasilinear problems, to show how the methods constructed for semilinear equations can be easily applied to more difficult problems.

On the contrary, Chapter 3 is still devoted to minimization techniques, but this time the examples are all taken from problems that lack compactness. Typically this may happen when the domain on which the equation is studied is not bounded or when the growth of the nonlinearity reaches a critical threshold. The chapter discusses some of the most common ways to overcome these problems, that are often very hard to solve. Of course only some prototype cases are presented, which is in the spirit of these notes.

Chapter 4 introduces the reader to the more refined variational methods, where minimization is replaced by minimax procedures. The methods are first introduced in an abstract setting, so that the reader can dispose of a general principle that is suited to handling more sophisticated procedures, beyond the scope of this book. Then we concentrate on the two most popular and widespread results: the Mountain Pass Theorem and the Saddle Point Theorem, each discussed with applications to the specific problems that motivated them.

Each chapter is accompanied by a selection of exercises, some of them guided, to test the reader's ability to refine and put into action the techniques just discussed.

A short bibliographical note closes each chapter: the interested reader can find there suggestions for further reading, mainly taken from more advanced books or from the celebrated papers that first solved the problems discussed in the text.

Torino, Italy   Marino Badiale
                                                                                                  Enrico Serra

Contents

Chapter 1
Introduction and Basic Results

In this chapter, after a short historical overview, we introduce the fundamental notions to be used later and the first existence results, based on convexity arguments.

1.1 Motivations and Brief Historical Notes

Semilinear elliptic equations are the first nonlinear generalization of linear elliptic partial differential equations. It is well known that linear elliptic equations, like the ubiquitous Laplace and Poisson equations, represent models for wide classes of physical problems. This is the reason why they have been studied for more than two hundred years and continue to attract researchers even today. One example is that the solutions of the Laplace and Poisson equations can represent the *stationary* solutions (i.e. the state of a system when equilibrium is reached) of evolution equations of the utmost physical importance, such as heat and wave equations. Solutions of these equations also represent or describe the potential of force fields, in a variety of physical contexts, like electromagnetism, gravitation, fluid dynamics and so on.

It is also noted that the linear equations commonly used in mathematical physics are very often just a first approximation, valid within the range of certain assumptions, of more complex equations modelling phenomena of a nonlinear nature. Under slightly less restrictive assumptions one is frequently led from linear to *semilinear* equations, namely equations where the nonlinearity involves the unknown function, but not its derivatives. In addition, we recall that semilinear elliptic equations manifest themselves even in quantum physics, despite the fact that the fundamental structure of quantum physics is of a linear nature.

Some examples of semilinear elliptic equations include

$$-\Delta u = f(x, u),$$

representing the stationary states of *nonlinear heat equations* like

$$u_t - \Delta u = f(x, u),$$

or *nonlinear wave equations*, such as

$$u_{tt} - \Delta u = f(x, u).$$

M. Badiale, E. Serra, *Semilinear Elliptic Equations for Beginners*, Universitext, DOI 10.1007/978-0-85729-227-8_1, © Springer-Verlag London Limited 2011

In all these examples u is the unknown, f is a (nonlinear) function, and Δ is the Laplace operator.

In optics and in the theory of water waves one encounters the *nonlinear Schrödinger equations*,

$$i u_t + \Delta u = \kappa |u|^p u,$$

and again the stationary solutions solve the corresponding semilinear equation.

The elliptic *sine-Gordon equation*

$$-\Delta u + \sin u = 0$$

appears in geometry, in the study of surfaces of constant negative curvature, and, again in geometry, important semilinear equations are related to classical problems, such as prescribing the scalar curvature on a manifold (the *Yamabe problem*).

The theory of Bose–Einstein condensates uses the time-independent *Gross–Pitaevskii equation*

$$-\Delta u + q(x)u + a|u|^2 u = \lambda u.$$

The *scalar field equations* in classical or quantum physics are noteworthy; they have the form

$$-\Delta u = f(u),$$

for example

$$-\Delta u = u^3 - u.$$

These are just a few examples of semilinear elliptic equations, which illustrate their frequent appearance in various areas of science.

The next step towards more complex nonlinear equations is given by *quasilinear* equations, where the nonlinear dependence involves not only the unknown function, but also all of its derivatives except those of the highest order. A very common class of quasilinear equations is given by the so-called *p*-Laplace equations, where the Laplacian is replaced by the operator

$$\operatorname{div}\left(|\nabla u|^{p-2}\nabla u\right),$$

which reduces to the usual Laplacian when $p = 2$.

The highest degree of nonlinearity is reached with *fully nonlinear* equations, in which the unknown function and *all* its derivatives are combined nonlinearly, as in the Monge–Ampère equations

$$\det\left(D^2 u\right) = f(x, u, \nabla u).$$

As a general principle, the more nonlinear the equation is, the more difficult it is.

In this book we concentrate on some types of semilinear elliptic equations, with some excursions into the realm of *p*-Laplace equations, confining ourselves to the cases where the theory developed for semilinear equations extends easily to the quasilinear context.

Of course elliptic equations can be studied with an astonishing variety of methods and techniques. In this book we present the *variational* approach, which is relatively simple, elegant and successful with large classes of problems.

Variational methods have a long history, which probably originated from the brachistochrone problem posed in 1696 and solved by Newton, Leibniz, Jakob and Johann Bernoulli, as well as de L'Hôpital, although the problem of surfaces of revolution in a resisting medium studied by Newton dates back even further. Major contributions were also given by Euler, who published the first monograph on the *Calculus of Variations* in 1744, and by Lagrange who introduced formalisms and techniques essentially still in use today. The Maupertuis least action principle was another of the first variational frameworks to be set.

The "classical" example for the problems studied in this book is the *Dirichlet Principle*, namely the fact that the solution of a differential equation (of elliptic type) coupled with a boundary condition can be obtained as a minimizer of an appropriate functional. In one of the most classical cases, an open set Ω in the plane or in the space is given, along with a real function f defined on the boundary of Ω. The problem consists in finding a function $u : \overline{\Omega} \to \mathbb{R}$ satisfying the boundary value problem

$$\begin{cases} \Delta u = 0 & \text{in } \Omega, \\ u = f & \text{on } \partial\Omega. \end{cases} \tag{1.1}$$

Under suitable assumptions, the (unique) solution of (1.1) is characterized as being the absolute minimum of the functional

$$I(u) = \int_{\Omega} |\nabla u|^2 \, dx$$

among all u's taking the value f on $\partial\Omega$ and belonging to a convenient function space.

Riemann introduced this point of view in 1851, and gave the name of Dirichlet Principle to the process of solving (1.1) by minimizing I. Although the Dirichlet Principle is elegant and can be applied to wide classes of problems other than (1.1), the proof supplied by Riemann cannot be considered correct, as was duly pointed out by Weierstrass. The problem lies in the fact that when working in function spaces, and hence in infinite dimensional vector spaces, the familiar compactness properties from finite dimension are no longer true, and the schemes of proof of existence of extrema, largely based on compactness arguments, break down. These facts had not yet been fully recognized at the time of Riemann.

The theoretical settlement of these kinds of difficulties required and stimulated the extraordinary development of functional analysis, measure theory and integration that exploded during the twentieth century, with the accurate study of topological and metric properties of infinite dimensional vector spaces, the Lebesgue theory of integration, and many other theoretical cornerstones.

The fundamental idea behind the Dirichlet Principle is the interpretation of a differential problem, written abstractly as $F(u) = 0$, as

$$I'(u) = 0,$$

where I is a suitable functional defined on a set of functions, and I' the differential of I in a sense to be made precise. In other words, zeros of F are seen as critical points (not necessarily minima) of I. The equation $I'(u) = 0$ is the *Euler*, or *Euler–Lagrange* equation associated with I.

Many times, it turns out that it is much easier to find a critical point of I than to work directly on the equation $F(u) = 0$. Furthermore, in countless applications the functional I has a fundamental physical meaning. Often I is an *energy* of some sort, written as the integral of a Lagrangian, and hence finding a minimum point means not only solving the differential equation, but finding the solution of minimal energy, frequently of particular relevance in concrete problems. The interpretation of I as an energy is so frequent that the functionals associated with differential problems are normally called *energy functionals*, even when the problem has no direct physical applications.

Of course not all differential problems can be written in the form $I'(u) = 0$. When this is possible, one says that the problem is *variational*, or has a variational structure, and these are the only type of problems that we address in these notes.

In the course of the twentieth century, after the settlement of functional analysis, the study of function spaces, the extension of differential calculus to normed spaces, variational methods have never ceased to be developed. On the one hand, minimization techniques have evolved to a very high level of efficiency, and have been applied to an enormous number of problems from all fields of pure and applied science. The body of methods concerned with minimization of functionals goes under the name of *direct methods* of the Calculus of Variations. On the other hand, the procedures aimed at the search for critical points of functionals that need not be minimum points have given rise to a branch of nonlinear analysis known as *Critical Point Theory*. Among the precursors of this theory are Ljusternik and Schnirelman, with their celebrated 1929 work on the existence of closed geodesics, and Morse, who laid the foundations of global analysis.

One of the most fruitful ideas that came out of this early research is the notion that the existence of critical points is very intimately related to the topological properties of the *sublevel* sets of the functional, in the sense that a change in the topological type of sublevels reveals the existence of a critical point, provided some compactness properties are satisfied. The systematization of the required compactness properties in the infinite dimensional setting is due to Palais and Smale, who introduced a compactness condition that bears their name and is nowadays accepted as the most functional notion.

The two factors—change of topology and compactness—have been later encompassed in specific, ready-to-use results for the researcher working in semilinear elliptic equations. The two most famous such results are the 1973 *Mountain Pass* Theorem by Ambrosetti and Rabinowitz, and the 1978 *Saddle Point* Theorem by Rabinowitz. This is where the "elementary" variational methods end, and make space for the most recent and sophisticated techniques, starting from linking theorems, index theories to prove the existence of multiple critical points, and onwards to the borders of current research.

1.2 Notation and Preliminaries

The notation we use throughout this book is standard. For convenience we list below the main symbols and notions that the reader is supposed to know. Proofs can be found in any text in Functional Analysis, such as [11], or [17].

The symbol Ω will always stand for an *open* subset of $\mathbb{R}^N$. The space $\mathbb{R}^N$ is endowed with the Lebesgue measure dx, and all integrals are to be considered in the sense of Lebesgue. The measure of a set $E \subset \mathbb{R}^N$ is denoted by $|E|$. We say that a property holds *almost everywhere* (shortly: a.e.) in Ω if there exists a set $E \subset \Omega$ of measure zero such that the property holds at every point of $\Omega \setminus E$.

The scalar product of two vectors x, y in $\mathbb{R}^N$ is denoted by a dot: $x \cdot y$, while the scalar product of two vectors u, v in a Hilbert space H is denoted by the symbol $(u \mid v)$.

If $u : \Omega \to \mathbb{R}$ is differentiable, the *gradient* of u at $x = (x_1, \ldots, x_N)$ is the vector $\nabla u(x) = (\frac{\partial u}{\partial x_1}(x), \ldots, \frac{\partial u}{\partial x_N}(x))$.

The letter C denotes a positive constant that can change from line to line. Finally, when taking limits, the symbol $o(1)$ stands for any quantity that tends to zero.

1.2.1 Function Spaces

- $C^k(\Omega)$, for $k = 1, 2, \ldots$ is the space of functions $u : \Omega \to \mathbb{R}$ that are k times differentiable in Ω and whose k-th derivatives are continuous in Ω.
- $C^\infty(\Omega)$ is the space of functions $u : \Omega \to \mathbb{R}$ that are infinitely many times differentiable in Ω.
- $C_0^\infty(\Omega)$ is the subspace of $C^\infty(\Omega)$ consisting of functions with *compact support* in Ω; the support of a (continuous) function $u : \Omega \to \mathbb{R}$, denoted by spt u is the *closure* (in $\mathbb{R}^N$) of the set $\{x \in \Omega \mid u(x) \neq 0\}$. Likewise, $C_0^k(\Omega)$ is the subset of $C^k(\Omega)$ containing only functions with compact support.
- $L^p(\Omega)$, for $p \in [1, +\infty)$, is the Lebesgue space of measurable functions (Lebesgue measure) $u : \Omega \to \mathbb{R}$ such that $\int_\Omega |u(x)|^p \, dx < +\infty$, while $L^\infty(\Omega)$ is the space of measurable functions such that ess $\sup_{x\in\Omega} |u(x)| < +\infty$. We recall that

$$\text{ess sup} |u(x)| = \inf\{C > 0 \mid |u(x)| \le C \text{ a.e. in } \Omega\}.$$

The norms that make $L^p(\Omega)$ Banach spaces are, respectively,

$$|u|_p = \left(\int_\Omega |u(x)|^p \, dx \right)^{1/p} \quad \text{and} \quad |u|_\infty = \text{ess sup} |u(x)|.$$

- $L_{\text{loc}}^p(\Omega)$ is the space of measurable functions $u : \Omega \to \mathbb{R}$ such that for every compact set $K \subset \Omega$, $|u|_{L^p(K)} < +\infty$.
- $H^1(\Omega)$ is the Sobolev space defined by

$$H^1(\Omega) = \left\{ u \in L^2(\Omega) \; \middle| \; \frac{\partial u}{\partial x_i} \in L^2(\Omega), i = 1, \ldots, N \right\},$$

where the derivative $\frac{\partial u}{\partial x_i}$ is in the sense of distributions. It is a Hilbert space, when endowed with the scalar product given by

$$(u \mid v) = \int_\Omega \nabla u \cdot \nabla v \, dx + \int_\Omega uv \, dx.$$

The corresponding norm is therefore

$$\|u\| = \sqrt{(u \mid u)} = \left(\int_\Omega |\nabla u|^2 \, dx + \int_\Omega u^2 \, dx \right)^{1/2}.$$

- $H_0^1(\Omega)$ is the closure of $C_0^\infty(\Omega)$ in $H^1(\Omega)$. The norm in $H^1(\Omega)$ and in $H_0^1(\Omega)$ will always be denoted by $\|u\|$.

  Functions in $H_0^1(\Omega)$ are to be thought of as functions that vanish on $\partial\Omega$. This is a delicate notion, since $\partial\Omega$ has measure zero, and a complete justification of the preceding sentence can only be carried out by means of the theory of traces (see e.g. [18]). However, as soon as u is regular enough to allow classical restrictions, there is a complete equivalence, in the following sense. Assume that $\partial\Omega$ is smooth, and that $u \in H^1(\Omega) \cap C(\overline\Omega)$. Then $u \in H_0^1(\Omega)$ if and only if $u = 0$ on $\partial\Omega$ (see [11]).

  A useful extension property of $H_0^1(\Omega)$ is the following. Let $u \in H_0^1(\Omega)$, and let Ω' be an open set such that $\Omega \subset \Omega'$. If one extends u to Ω' by setting

$$\tilde{u}(x) = \begin{cases} u(x) & \text{if } \in \Omega, \\ 0 & \text{if } x \in \Omega' \setminus \Omega, \end{cases}$$

then $\tilde{u} \in H_0^1(\Omega')$.

  The space $H_0^1(\mathbb{R}^N)$ coincides with $H^1(\mathbb{R}^N)$.

  Finally, we recall (see [18]) that if $u \in H^1(\Omega)$, then $|u| \in H^1(\Omega)$; therefore if we set

$$u^+(x) = \max(u(x), 0) \quad \text{and} \quad u^-(x) = -\min(u(x), 0),$$

then $u^+, u^- \in H^1(\Omega)$, since $u^+ = \frac{1}{2}(u + |u|)$ and $u^- = \frac{1}{2}(|u| - u)$.

  Also,

$$\int_\Omega |\nabla|u||^2 \, dx = \int_\Omega |\nabla u|^2 \, dx$$

for every $u \in H^1(\Omega)$.

- $D^{1,2}(\mathbb{R}^N)$, for $N \geq 3$, is the space defined as follows: let $2^* = \frac{2N}{N-2}$ (see the next subsection for a short description of the role of this number); then

$$D^{1,2}(\mathbb{R}^N) = \left\{ u \in L^{2^*}(\mathbb{R}^N) \mid \frac{\partial u}{\partial x_i} \in L^2(\mathbb{R}^N), i = 1, 2, \ldots, N \right\}.$$

This space has a Hilbert structure when endowed with the scalar product

$$(u \mid v) = \int_{\mathbb{R}^N} \nabla u \cdot \nabla v \, dx,$$

so that the corresponding norm is

$$\|u\| = \sqrt{(u \mid u)} = \left(\int_{\mathbb{R}^N} |\nabla u|^2 \, dx \right)^{1/2}.$$

The space $C_0^\infty(\mathbb{R}^N)$ is dense in $D^{1,2}(\mathbb{R}^N)$.

Of course $H^1(\mathbb{R}^N) \subset D^{1,2}(\mathbb{R}^N)$, but there are functions, such as

$$u(x) = \frac{1}{(1 + |x|)^{N/2}},$$

that are in $D^{1,2}(\mathbb{R}^N)$ but not in $L^2(\mathbb{R}^N)$, and hence not in $H^1(\mathbb{R}^N)$.

1.2.2 Embeddings

We recall that a Banach space X is embedded *continuously* in a Banach space Y ($X \hookrightarrow Y$) if

1. $X \subseteq Y$;
2. the canonical injection $j : X \to Y$ is a continuous (linear) operator. This means that there exists a constant $C > 0$ such that $\|j(u)\|_Y \leq C\|u\|_X$, which one writes $\|u\|_Y \leq C\|u\|_X$, for every $u \in X$.

A Banach space X is embedded *compactly* in a Banach space Y if X is embedded continuously in Y and the canonical injection j is a compact operator.

The following results are the particular cases of the Sobolev and Rellich Embedding theorems that we need in this book. First we deal with functions defined on bounded sets.

Theorem 1.2.1 *Let $\Omega \subset \mathbb{R}^N$ be an open and bounded subset of $\mathbb{R}^N$, with $N \geq 3$. Then*

$$H_0^1(\Omega) \hookrightarrow L^q(\Omega) \quad \text{for every } q \in \left[1, \frac{2N}{N-2}\right].$$

The embedding is compact if and only if $q \in [1, \frac{2N}{N-2})$.

The number $\frac{2N}{N-2}$ is denoted by 2^* and is called the *critical Sobolev exponent* for the embedding of H_0^1 into L^q. The term "critical" refers to the fact that the embedding of the preceding theorem fails for $q > 2^*$.

For functions defined on general, unbounded domains, in view of our applications we limit ourselves to the case $\Omega = \mathbb{R}^N$.

Theorem 1.2.2 *Let $N \geq 3$. Then*

- $H^1(\mathbb{R}^N) \hookrightarrow L^q(\mathbb{R}^N)$ *for every $q \in [2, \frac{2N}{N-2}]$;*
- $D^{1,2}(\mathbb{R}^N) \hookrightarrow L^{2^*}(\mathbb{R}^N)$.

These embeddings are never compact.

We point out that the continuity of the above embeddings is expressed explicitly by inequalities of the form

$$|u|_q \leq C\|u\| \quad \forall u \in H_0^1(\Omega),$$

where C does not depend on u. Inequalities of this type are often referred to as *Sobolev inequalities*.

1.2.3 Elliptic Equations

This book is devoted to *elliptic* differential equations of the second order. We recall the definition of elliptic differential operator.

Let Ω be an open subset of $\mathbb{R}^N$, with $N \geq 2$. A second order linear differential operator L with nonconstant coefficients acting on functions $u : \Omega \to \mathbb{R}$ has the general form

$$Lu = -\sum_{i,j=1}^N a_{ij}(x)\frac{\partial^2 u}{\partial x_i \partial x_j} + \sum_{i=1}^N b_i(x)\frac{\partial u}{\partial x_i} + q(x)u, \tag{1.2}$$

where the coefficients a_{ij}, b_i and q are functions defined on Ω satisfying some regularity requirement, such as continuity, boundedness, measurability, or other assumptions that depend on the problem one is interested in.

The operator L is said to be in *divergence form* if

$$Lu = -\sum_{i,j=1}^N \frac{\partial}{\partial x_i}\left(a_{ij}(x)\frac{\partial u}{\partial x_j}\right) + \sum_{i=1}^N b_i(x)\frac{\partial u}{\partial x_i} + q(x)u. \tag{1.3}$$

Every operator in the form (1.2) can be put in divergence form, and conversely, every operator in divergence form can be written as in (1.2), *provided* the coefficients a_{ij} are smooth enough (say differentiable). Passing from one form to the other alters the coefficients of the first order terms, but nothing else. Hence, for smooth coefficients, the two forms can be considered equivalent.

The name "divergence form" has the following explanation. Define a function A from Ω to the set of $N \times N$ matrices by

$$A(x) = (a_{ij}(x))_{ij}.$$

Then the principal part of the operator (namely the part containing the second order derivatives) is exactly

$$-\operatorname{div}(A(x)\nabla u).$$

Definition 1.2.3   A second order linear differential operator L given by (1.2) or (1.3) is called *uniformly elliptic* on Ω if there exists $\lambda > 0$ such that

$$\sum_{i,j=1}^N a_{ij}(x)\xi_i\xi_j \geq \lambda|\xi|^2 \quad \forall \xi \in \mathbb{R}^N \text{ and for a.e. } x \in \Omega.$$

In other words, L is (uniformly) elliptic if the quadratic form $\xi \mapsto A(x)\xi \cdot \xi$ is (uniformly) positive definite in Ω.

Uniformly elliptic operators tend to behave all in the same way for many questions. However, for the *variational* approach, which is the object of this book, the first order terms have to be neglected. Therefore we will always consider operators in divergence form without first order terms, namely

$$Lu = -\operatorname{div}(A(x)\nabla u) + q(x)u,$$

and we will always assume L to be uniformly elliptic. Since under reasonable assumptions on the matrix A these operators really behave in the same way, it is customary to develop the theory for the model case, in which A is the identity matrix I. In this case,

$$Lu = -\operatorname{div}(A(x)\nabla u) = -\operatorname{div}(I\nabla u) = \sum_{i=1}^{N} \frac{\partial^2 u}{\partial x_i^2} =: -\Delta u. \tag{1.4}$$

The operator Δ is called the *Laplace operator*, or the *Laplacian*.

Remark 1.2.4 The reader may perhaps wonder why the differential operators are written with a "minus" sign in front of them. The reason will be clear in Sect. 1.4: in the definition of weak solution the minus sign disappears, and it is preferable to write it in the equation than to have it the definition of weak solution or in other quantities of common use.

1.2.4 Frequently Used Results

As a fast reference for the reader we state some classical results that we will use freely in the sequel.

Theorem 1.2.5 (Green's formula) *Let $\Omega \subset \mathbb{R}^N$ be open, bounded and smooth. Let $u \in C^2(\overline{\Omega})$ and $v \in C^1(\overline{\Omega})$. Then*

$$\int_{\Omega} (\Delta u)\, v\, dx = \int_{\partial\Omega} \frac{\partial u}{\partial \nu}\, v\, d\sigma - \int_{\Omega} \nabla u \cdot \nabla v\, dx, \tag{1.5}$$

where $\nu = \nu(x)$ is the outward normal to $\partial\Omega$ at x, $\frac{\partial u}{\partial \nu}(x) = \nabla u(x) \cdot \nu(x)$ and σ is the surface measure on $\partial\Omega$.

Theorem 1.2.6 (Lebesgue, or dominated convergence) *Let $\Omega \subset \mathbb{R}^N$ be open and let $\{u_k\}_k \subset L^1(\Omega)$ be a sequence such that*

1. *$u_k(x) \to u(x)$ a.e. in Ω as $k \to \infty$;*
2. *there exists $v \in L^1(\Omega)$ such that for all k, $|u_k(x)| \le v(x)$ a.e. in Ω.*

Then $u \in L^1(\Omega)$ and $u_k \to u$ in the $L^1(\Omega)$ norm, namely $\int_{\Omega} |u_k - u|\, dx \to 0$.

Theorem 1.2.7 *Let $\Omega \subset \mathbb{R}^N$ be open and let $\{u_k\}_k \subset L^p(\Omega)$, $p \in [1, +\infty]$, be a sequence such that $u_k \to u$ in $L^p(\Omega)$ as $k \to \infty$. Then there exist a subsequence $\{u_{k_j}\}_j$ and a function $v \in L^p(\Omega)$ such that*

1. $u_{k_j}(x) \to u(x)$ *a.e. in Ω as $j \to \infty$;*
2. *for all j, $|u_{k_j}(x)| \leq v(x)$ a.e. in Ω.*

Theorem 1.2.8 (Poincaré inequality) *Let $\Omega \subset \mathbb{R}^N$ be open and bounded. Then there exists a constant $C > 0$, depending only on Ω, such that*

$$\int_\Omega u^2 \, dx \leq C \int_\Omega |\nabla u|^2 \, dx \quad \forall u \in H_0^1(\Omega).$$

As a consequence, the quantity $(\int_\Omega |\nabla u|^2 \, dx)^{\frac{1}{2}}$ is a norm on $H_0^1(\Omega)$, equivalent to the standard one.

Theorem 1.2.9 (Riesz) *Let H be a Hilbert space, and let H' be its topological dual. Then for every $f \in H'$ there exists a unique $u_f \in H$ such that*

$$f(v) = (u_f \mid v) \quad \forall v \in H.$$

Moreover, $\|u_f\|_H = \|f\|_{H'}$. The linear application $R : H' \to H$ that sends f to u_f is called the Riesz isomorphism.

Theorem 1.2.10 (Banach–Alaoglu) *Let X be a reflexive Banach space. If $B \subset X$ is bounded, then B is relatively compact in the weak topology of X.*

In a Banach space X with dual X', we write $u_k \to u$ when the sequence $\{u_k\}_k$ converges *strongly* to u, that is, in the *strong* topology of X, which means that $\|u_k - u\|_X \to 0$ as $k \to \infty$; we write $u_k \rightharpoonup u$ if u_k converges *weakly* to u, i.e. in the *weak* topology of X, which means that

$$f(u_k) \to f(u) \quad \text{as } k \to \infty \, \forall f \in X'.$$

Example 1.2.11 The following chain of arguments is used very frequently, often in an automatic way. Let $\Omega \subset \mathbb{R}^N$ be open and bounded. Suppose that a sequence $\{u_k\}_k \subset H_0^1(\Omega)$ satisfies

$$\int_\Omega |\nabla u_k|^2 \, dx \leq C$$

for all k, for some $C > 0$ independent of k. By Theorem 1.2.8, the sequence u_k is bounded in $H_0^1(\Omega)$. The space $H_0^1(\Omega)$, being a Hilbert space, is reflexive. Therefore by the Banach–Alaoglu Theorem the sequence is relatively compact in $H_0^1(\Omega)$

endowed with the weak topology. This means that there exist $u \in H_0^1(\Omega)$ and a subsequence that we call again u_k, such that

$$u_k \rightharpoonup u \quad \text{in } H_0^1(\Omega).$$

By Theorem 1.2.1, the embedding of $H_0^1(\Omega)$ into $L^p(\Omega)$ is compact for every $p \in [1, \frac{2N}{N-2})$. Then we can say that

$$u_k \to u \quad \text{in } L^p(\Omega) \text{ for every } p \in \left[1, \frac{2N}{N-2}\right).$$

By Theorem 1.2.7 there exists another subsequence, still denoted u_k, such that also

$$u_k(x) \to u(x) \quad \text{a.e. in } \Omega.$$

Summing up, whenever a sequence $\{u_k\}_k \subset H_0^1(\Omega)$ satisfies $\sup_k \int_\Omega |\nabla u_k|^2 \, dx < +\infty$, we can deduce that there exists $u \in H_0^1(\Omega)$ such that, up to subsequences,

- $u_k \rightharpoonup u$ in $H_0^1(\Omega)$;
- $u_k \to u$ in $L^p(\Omega), \forall p \in [1, \frac{2N}{N-2})$;
- $u_k(x) \to u(x)$ a.e. in Ω;

moreover, again by Theorem 1.2.7, there exists $w \in L^p(\Omega)$ such that $|u_k(x)| \le w(x)$ a.e. in Ω and for all k.

1.3   A Review of Differential Calculus for Real Functionals

We present a short review of the main definitions and results concerning the differential calculus for real functionals defined on a Banach space. A complete discussion of this subject, and more generally of differential calculus in normed spaces can be found for example in [4].

Whenever X is a Banach space, we denote by X' its (topological) *dual*, namely the space of continuous *linear* functionals from X to $\mathbb{R}$. We recall that X' is a Banach space endowed with the norm

$$\|A\| = \sup_{\substack{u \in X \\ \|u\|=1}} |A(u)|.$$

If X is a Banach space and $U \subset X$, a *functional* I on U is an application $I : U \to \mathbb{R}$. The elements of X' are particular cases of functionals (they are linear and continuous). In this book the attention is devoted to general, nonlinear functionals.

We present the two principal definitions of differentiability and their main properties. Examples will be given next.

Definition 1.3.1 Let X be a Banach space, U an open subset of X and let $I : U \to \mathbb{R}$ be a functional. We say that I is (Fréchet) differentiable at $u \in U$ if there exists $A \in X'$ such that

$$\lim_{\|v\| \to 0} \frac{I(u+v) - I(u) - Av}{\|v\|} = 0. \tag{1.6}$$

Thus, for a differentiable functional I, we have

$$I(u + v) - I(u) = Av + o(\|v\|)$$

as $\|v\| \to 0$ for some $A \in X'$, namely the increment $I(u + v) - I(u)$ is *linear* in v, up to a higher order quantity. This of course implies that if I is differentiable at u, then I is continuous at u.

Remark 1.3.2 If a functional I is differentiable at u, there is a unique $A \in X'$ satisfying Definition 1.3.1. Indeed, if A and B are two elements of X' that satisfy (1.6), then plainly

$$\lim_{\|v\| \to 0} \frac{(A - B)v}{\|v\|} = 0,$$

so that, if $u \in X$ and $\|u\| = 1$,

$$(A - B)u = \lim_{t \to 0^+} \frac{(A - B)(tu)}{t} = 0,$$

which means $A = B$.

Definition 1.3.3 Let $I : U \to \mathbb{R}$ be differentiable at $u \in U$. The unique element of X' such that (1.6) holds is called the (Fréchet) differential of I at u, and is denoted by $I'(u)$ or by $dI(u)$. We thus have

$$I(u + v) = I(u) + I'(u)v + o(\|v\|) \tag{1.7}$$

as $\|v\| \to 0$.

Notice that $I'(u)$ is defined on X, even if I is defined only in U.

In the literature the terms differentiability and Fréchet differentiability are commonly interchanged. We will also do so, unless we deal with more than one notion of differentiability at the same time.

Definition 1.3.4 Let $U \subseteq X$ be an open set. If the functional I is differentiable at every $u \in U$, we say that I is differentiable on U. The map $I' : U \to X'$ that sends $u \in U$ to $I'(u) \in X'$ is called the (Fréchet) derivative of I. Note that I' is in general a *nonlinear* map. If the derivative I' is continuous from U to X' we say that I is of class C^1 on U and we write $I \in C^1(U)$.

A particular but very important case is that of real functionals defined on a *Hilbert* space H with scalar product $(\cdot|\cdot)$. Indeed since H' and H can be identified via the *Riesz isomorphism* $R : H' \to H$ (see Theorem 1.2.9), the linear functionals on H can be represented by the scalar product in H, in the sense that for every $A \in H'$ there exists a unique $RA \in H$ such that

$$A(u) = (RA|u) \quad \text{for every } u \in H.$$

Definition 1.3.5 Let H be a Hilbert space, $U \subseteq H$ an open set and let $R : H' \to H$ be the Riesz isomorphism. Assume that the functional $I : U \to \mathbb{R}$ is differentiable at u. The element $RI'(u) \in H$ is called the gradient of I at u and is denoted by $\nabla I(u)$; therefore

$$I'(u)v = (\nabla I(u)|v) \quad \text{for every } v \in H.$$

The following proposition collects the properties of differentiable functionals that we will use repeatedly.

Proposition 1.3.6 *Assume that I and J are differentiable at $u \in U \subseteq X$. Then the following properties hold:*

1. *if a and b are real numbers, $aI + bJ$ is differentiable at u and*

$$(aI + bJ)'(u) = aI'(u) + bJ'(u);$$

2. *the product IJ is differentiable at u and*

$$(IJ)'(u) = J(u)I'(u) + I(u)J'(u);$$

3. *if $\gamma : \mathbb{R} \to U$ is differentiable at t_0 and $u = \gamma(t_0)$, then the composition $\eta : \mathbb{R} \to \mathbb{R}$ defined by $\eta(t) = I(\gamma(t))$ is differentiable at t_0 and*

$$\eta'(t_0) = I'(u)\gamma'(t_0);$$

4. *if $A \subseteq \mathbb{R}$ is an open set, $f : A \to \mathbb{R}$ is differentiable at $I(u) \in A$, then the composition $K(u) = f(I(u))$ is defined in an open neighborhood V of u, is differentiable at u and*

$$K'(u) = f'(I(u))I'(u).$$

Proof The first statement is trivial. As to the second, when $v \to 0$ in X, we have

$$I(u + v)J(u + v) = \Big(I(u) + I'(u)v + o(\|v\|)\Big)\Big(J(u) + J'(u)v + o(\|v\|)\Big)$$

$$= I(u)J(u) + J(u)I'(u)v + I(u)J'(u)v$$

$$+ I'(u)vJ'(u)v + o(\|v\|)$$

$$\times \Big(I(u) + I'(u)v + J(u) + J'(u)v + o(\|v\|)\Big).$$

Since the last line is $o(\|v\|)$, we obtain

$$I(u + v)J(u + v) = I(u)J(u) + \Big(J(u)I'(u) + I(u)J'(u)\Big)v + o(\|v\|),$$

as we wanted to prove.
   In a similar way, as $h \to 0$,

$$\eta(t_0 + h) = I(\gamma(t_0 + h)) = I(\gamma(t_0) + \gamma'(t_0)h + o(h))$$

$$= I(u) + I'(u)(\gamma'(t_0)h + o(h)) + o(\|\gamma'(t_0)h + o(h)\|)$$

$$= \eta(t_0) + I'(u)\gamma'(t_0)h + I'(u)o(h) + o(\|\gamma'(t_0)h + o(h)\|).$$

Since the last two terms are $o(|h|)$, we obtain

$$\eta(t_0 + h) = \eta(t_0) + (I'(u)\gamma'(t_0))h + o(|h|),$$

namely η is differentiable at t_0 and $\eta'(t_0) = I'(u)\gamma'(t_0)$.

Finally, when $v \to 0$ in X, one easily checks that

$$K(u + v) = f(I(u + v)) = f(I(u) + I'(u)v + o(\|v\|))$$

$$= f(I(u)) + f'(I(u))(I'(u)v + o(\|v\|)) + o(I'(u)v + o(\|v\|))$$

$$= f(I(u)) + f'(I(u))I'(u)v + o(\|v\|),$$

so that also the fourth statement follows.   □

We now introduce a second, weaker notion of differentiability. This is often simpler to check in concrete cases than (Fréchet) differentiability.

Definition 1.3.7 Let X be a Banach space, $U \subseteq X$ an open set and let $I : U \to \mathbb{R}$ be a functional. We say that I is Gâteaux differentiable at $u \in U$ if there exists $A \in X'$ such that, for all $v \in X$,

$$\lim_{t \to 0} \frac{I(u + tv) - I(u)}{t} = Av. \tag{1.8}$$

If I is Gâteaux differentiable at u, there is only one linear functional $A \in X'$ satisfying (1.8). It is called the Gâteaux differential of I at u and is denoted by $I'_G(u)$.

By the very definition of Fréchet differentiability, it is obvious that if I is differentiable at u, then it is also Gâteaux differentiable and $I'(u) = I'_G(u)$. It is not true that Gâteaux differentiability implies differentiability, exactly as in $\mathbb{R}^N$ directional differentiability does not imply differentiability. However, Proposition 1.3.8 below gives a relevant result in this direction.

As for the notion of differentiability, if the functional I is Gâteaux differentiable at every u of an open set $U \subset X$, we say that I is Gâteaux differentiable on U. The (generally nonlinear) map $I'_G : U \to X'$ that sends $u \in U$ to $I'_G(u) \in X'$ is called the Gâteaux derivative of I.

We have the following classical result; for the proof see e.g. [4].

Proposition 1.3.8 *Assume that $U \subseteq X$ is an open set, that I is Gâteaux differentiable on U and that I'_G is continuous at $u \in U$. Then I is also differentiable at u, and of course $I'_G(u) = I'(u)$.*

Remark 1.3.9 The importance of this proposition lies in the fact that it is often technically easier to compute the Gâteaux derivative and then prove that it is continuous, rather than proving directly the (Fréchet) differentiability.

We conclude with the definitions of critical points and critical levels, which will be one of the main concerns of this book.

Definition 1.3.10 Let X be a Banach space, $U \subseteq X$ an open set and assume that $I : U \to \mathbb{R}$ is differentiable. A critical point of I is a point $u \in U$ such that

$$I'(u) = 0.$$

As $I'(u)$ is an element of the dual space X', this means of course $I'(u)v = 0$ for all $v \in X$.

If $I'(u) = 0$ and $I(u) = c$, we say that u is a critical point for I at level c. If for some $c \in \mathbb{R}$ the set $I^{-1}(c) \subset X$ contains at least a critical point, we say that c is a critical level for I.

The equation $I'(u) = 0$ is called the Euler, or Euler–Lagrange equation associated to the functional I.

1.3.1   Examples in Abstract Spaces

We begin with some abstract examples of differentiable functionals.

Example 1.3.11 The simplest example of differentiable functional is given by a constant functional, which is everywhere differentiable with differential equal to zero. Another trivial example is given by a linear and continuous functional $A : X \to \mathbb{R}$, namely $A \in X'$. The functional A is everywhere differentiable and the differential at any point is A itself, that is

$$A'(u) = A$$

for every $u \in X$.

Example 1.3.12 A more interesting case is the following. Let X be a Banach space and let

$$a : X \times X \to \mathbb{R}$$

be a continuous bilinear form. Denote by $J : X \to \mathbb{R}$ the associated quadratic form given by

$$J(u) = a(u, u).$$

Then J is differentiable on X and

$$J'(u)v = a(u, v) + a(v, u), \quad \text{or} \quad J'(u) = a(u, \cdot) + a(\cdot, u)$$

for all $u, v \in X$. Indeed by linearity

$$J(u + v) - J(u) = a(u + v, u + v) - a(u, u) = a(u, v) + a(v, u) + a(v, v);$$

by continuity, $a(v, v) = o(\|v\|)$, and the claim follows.

Of course, if a is also *symmetric*, then

$$J'(u)v = 2a(u, v).$$

Example 1.3.13 A particular case occurs with a continuous bilinear form

$$a : H \times H \to \mathbb{R}$$

where H is a *Hilbert* space. Setting again $J(u) = a(u, u)$, the results described in the previous example apply. In particular, if a is symmetric, then $J'(u) = 2a(u, \cdot)$. Taking as a the scalar product, we have that

$$J(u) = (u|u) = \|u\|^2.$$

We obtain that on a Hilbert space, the square of the norm is everywhere differentiable and

$$J'(u) = 2(u|\cdot).$$

Since H is Hilbert, we can identify continuous linear functionals on H with elements of the space H itself, as we have already recalled. In this case therefore we can compute the gradient of J, which takes the particularly simple form

$$\nabla J(u) = 2u.$$

Example 1.3.14 Still on a Hilbert space H, one can easily prove that the functional $J(u) = \|u\|$ is differentiable at every $u \neq 0$, with

$$J'(u) = \left(\frac{u}{\|u\|} \Big| \cdot \right), \quad \text{or} \quad \nabla J(u) = \frac{u}{\|u\|}.$$

One can also prove that J is *not* differentiable at zero (see e.g. [4]).

Remark 1.3.15 The question of the differentiability of $\|u\|$ in absence of the Hilbert structure is much more delicate and depends in general on finer properties of the space, such as uniform convexity, reflexivity and so on. We will see some examples of differentiable norms in Sect. 2.6.

Example 1.3.16 We will sometimes deal with "quotient" functionals constructed in the following way: assume that I and J are two differentiable functionals on a Banach space X and define

$$Q(u) = \frac{I(u)}{J(u)}.$$

Of course Q is defined on $U = \{u \in X \mid J(u) \neq 0\}$. By Proposition 1.3.6 we have that Q is differentiable on U and

$$Q'(u)v = \frac{J(u)I'(u)v - I(u)J'(u)v}{J(u)^2} = \frac{1}{J(u)} \left(I'(u)v - Q(u)J'(u)v \right).$$

1.3.2 Examples in Concrete Spaces

The examples collected in this section will all be used in later chapters since they present the differentiability properties of functionals associated to elliptic equations.

We begin with a direct application of the "abstract" examples.

Example 1.3.17 Let $\Omega \subset \mathbb{R}^N$, $N \geq 1$, be a bounded open set. We define the functionals

$$I : L^2(\Omega) \to \mathbb{R} \quad \text{as } I(u) = \int_\Omega u(x)^2 \, dx,$$

$$J : H_0^1(\Omega) \to \mathbb{R} \quad \text{as } J(u) = \int_\Omega |\nabla u(x)|^2 \, dx,$$

$$K : H^1(\Omega) \to \mathbb{R} \quad \text{as } K(u) = \int_\Omega |\nabla u(x)|^2 \, dx,$$

$$L : H^1(\Omega) \to \mathbb{R} \quad \text{as } L(u) = \int_\Omega |\nabla u(x)|^2 \, dx + \int_\Omega u(x)^2 \, dx.$$

The functionals I, J, K and L are all quadratic forms associated to symmetric continuous bilinear forms. By the discussion in Example 1.3.13 they are differentiable at every point.

The functional I is the square of the norm in $L^2(\Omega)$ and L is the square of the norm in $H^1(\Omega)$; thus

$$I'(u)v = 2(u|v)_{L^2(\Omega)} = 2\int_\Omega u(x)v(x)\,dx \quad \text{and} \quad \nabla I(u) = 2u.$$

$$L'(u)v = 2(u|v)_{H^1(\Omega)} = 2\int_\Omega \nabla u(x) \cdot \nabla v(x)\,dx + 2\int_\Omega u(x)v(x)\,dx \quad \text{and}$$

$$\nabla L(u) = 2u.$$

The functional J is again the square of the norm, provided one uses in $H_0^1(\Omega)$ the norm $J(u)^{1/2}$, equivalent to the one induced by $H^1(\Omega)$ (see Theorem 1.2.8). Accordingly,

$$J'(u)v = 2(u|v)_{H_0^1(\Omega)} = 2\int_\Omega \nabla u(x) \cdot \nabla v(x)\,dx \quad \text{and} \quad \nabla J(u) = 2u.$$

The functional K is *not* the square of the norm in $H^1(\Omega)$; therefore

$$K'(u)v = 2\int_\Omega \nabla u(x) \cdot \nabla v(x)\,dx,$$

but we cannot say that $\nabla K(u) = 2u$.

Since $H_0^1(\Omega)$ and $H^1(\Omega)$ are both embedded continuously in $L^2(\Omega)$, quantities that are $o(|v|_2)$ as $v \to 0$ in $L^2(\Omega)$ are also $o(\|v\|_{H_0^1(\Omega)})$ or $o(\|v\|_{H^1(\Omega)})$ as $v \to 0$ in the respective topologies. Therefore I is also differentiable at every point of $H_0^1(\Omega)$ or of $H^1(\Omega)$. The expression of $I'(u)$ of course is unchanged.

Example 1.3.18 A slightly more general situation is the following: consider functions $a_{ij} \in L^\infty(\Omega)$ for $i, j \in \{1, \dots, N\}$, and define the $N \times N$ matrix $A(x) = (a_{ij}(x))_{ij}$. Take a function $q \in L^\infty(\Omega)$ and define a map $a : H^1(\Omega) \times H^1(\Omega) \to \mathbb{R}$ as

$$a(u, v) = \int_\Omega (A(x)\nabla u(x)) \cdot \nabla v(x)\,dx + \int_\Omega q(x)u(x)v(x)\,dx.$$

It is easy to see that a is a continuous bilinear form and that if $A(x)$ is a symmetric matrix, then a is also symmetric. Hence if we define

$$J(u) = a(u, u) = \int_\Omega (A(x)\nabla u(x)) \cdot \nabla u(x)\, dx + \int_\Omega q(x) u^2\, dx,$$

then J is differentiable on $H^1(\Omega)$ and

$$J'(u)v = 2a(u, v) = 2\int_\Omega (A(x)\nabla u(x)) \cdot \nabla v(x)\, dx + 2\int_\Omega q(x) u(x) v(x)\, dx.$$

Example 1.3.19 With the same assumptions and notation as in the previous example, define $Q : H^1(\Omega) \setminus \{0\} \to \mathbb{R}$ as

$$Q(u) = \frac{J(u)}{\int_\Omega u^2\, dx} = \frac{\int_\Omega (A(x)\nabla u) \cdot \nabla u\, dx + \int_\Omega q(x) u^2\, dx}{\int_\Omega u^2\, dx}.$$

Then, as in Example 1.3.16, Q is differentiable on $H^1(\Omega) \setminus \{0\}$ and there results

$$Q'(u)v = \frac{1}{\int_\Omega u^2\, dx} \left(\int_\Omega (A(x)\nabla u) \cdot \nabla v\, dx + \int_\Omega q(x) uv\, dx - Q(u)\int_\Omega uv\, dx \right).$$

We now turn to the differentiability properties of integral functionals like $u \mapsto \int_\Omega F(u)\, dx$. In dealing with these functionals we will make great use of the elementary inequality

$$\forall q > 0 \ \exists C_q \quad \text{such that} \quad |a + b|^q \le C_q \big(|a|^q + |b|^q\big), \forall a, b \in \mathbb{R}.$$

Example 1.3.20 Let $\Omega \subset \mathbb{R}^N$, $N \ge 3$, be a bounded open set (with smooth boundary). Let $f : \mathbb{R} \to \mathbb{R}$ be a continuous function, and assume that there exist $a, b > 0$ such that

$$|f(t)| \le a + b|t|^{2^*-1} \tag{1.9}$$

for all $t \in \mathbb{R}$. Recall that $2^* = \frac{2N}{N-2}$ is the critical exponent for the embedding of $H^1(\Omega)$ into $L^p(\Omega)$. Define

$$F(t) = \int_0^t f(s)\, ds$$

and consider the functional $J : H^1(\Omega) \to \mathbb{R}$ given by

$$J(u) = \int_\Omega F(u(x))\, dx.$$

Then J is differentiable on $H^1(\Omega)$ and

$$J'(u)v = \int_\Omega f(u(x)) v(x)\, dx \quad \text{for all } u, v \in H^1(\Omega).$$

The functional J can also be considered on $H_0^1(\Omega)$, without any regularity assumption on $\partial\Omega$, and the same result holds.

To check this we use a common procedure in this kind of questions: *first* we prove that J is Gâteaux differentiable, and *then* we show that J'_G is continuous, so that we can invoke Proposition 1.3.8.

From the growth assumption (1.9) one easily derives that $J(u)$ is well defined via the Sobolev inequalities (Theorem 1.2.1). We now check that J is Gâteaux differentiable. We have to prove that for fixed $u, v \in H^1(\Omega)$,

$$\lim_{t \to 0} \int_\Omega \frac{F(u + tv) - F(u)}{t}\, dx = \int_\Omega f(u)v\, dx.$$

It is obvious that, for almost every $x \in \Omega$,

$$\lim_{t \to 0} \frac{F(u(x) + tv(x)) - F(u(x))}{t} = f(u(x))v(x).$$

By the Lagrange Theorem there exists a real number θ such that $|\theta| \le |t|$ and

$$\left| \frac{F(u(x) + tv(x)) - F(u(x))}{t} \right| = \left| f(u(x) + \theta v(x))v(x) \right|$$

$$\le \left(a + b|u(x) + \theta v(x)|^{2^*-1} \right) |v(x)|$$

$$\le C \left(|v(x)| + |u(x)|^{2^*-1}|v(x)| + |v(x)|^{2^*} \right).$$

As the function $|v| + |u|^{2^*-1}|v| + |v|^{2^*}$ is in $L^1(\Omega)$, by dominated convergence we have

$$\lim_{t \to 0} \int_\Omega \frac{F(u + tv) - F(u)}{t}\, dx = \int_\Omega f(u)v\, dx.$$

Since the right-hand side, as a function of v, is a continuous linear functional on $H^1(\Omega)$, it is the Gâteaux differential of J.

We complete the proof by checking that the function $J'_G : H^1(\Omega) \to [H^1(\Omega)]'$ is continuous. To this aim, take a sequence $\{u_k\}_k$ in $H^1(\Omega)$ such that $u_k \to u$ in $H^1(\Omega)$. Up to a subsequence, we may assume that

- $u_k \to u$ in $L^{2^*}(\Omega)$ as $k \to \infty$;
- $u_k(x) \to u(x)$ a.e. in Ω as $k \to \infty$;
- there is $w \in L^{2^*}(\Omega)$ such that $|u_k(x)| \le w(x)$ a.e. in Ω and for all $k \in \mathbb{N}$.

We have, by the Hölder inequality,

$$\left| \left(J'_G(u_k) - J'_G(u) \right) v \right| \le \int_\Omega |f(u_k) - f(u)||v|\, dx$$

$$\le \left(\int_\Omega |f(u_k) - f(u)|^{\frac{2^*}{2^*-1}}\, dx \right)^{\frac{2^*-1}{2^*}} \left(\int_\Omega |v|^{2^*}\, dx \right)^{1/2^*}.$$

Since $\lim_{k \to +\infty} |f(u_k(x)) - f(u(x))| = 0$ a.e. in Ω and

$$|f(u_k) - f(u)|^{\frac{2^*}{2^*-1}} \leq C \left(1 + |u_k|^{2^*-1} + |u|^{2^*-1}\right)^{\frac{2^*}{2^*-1}}$$

$$\leq C \left(1 + |w|^{2^*-1} + |u|^{2^*-1}\right)^{\frac{2^*}{2^*-1}}$$

$$\leq C \left(1 + |w|^{2^*} + |u|^{2^*}\right) \in L^1(\Omega),$$

then by dominated convergence,

$$\lim_{k \to +\infty} \int_\Omega |f(u_k) - f(u)|^{\frac{2^*}{2^*-1}} \, dx = 0,$$

so that

$$\|J'_G(u_k) - J'_G(u)\| = \sup\left\{|(J'_G(u_k) - J'_G(u))v| \mid v \in H^1(\Omega), \|v\| = 1\right\}$$

$$\leq C \left(\int_\Omega |f(u_k) - f(u)|^{\frac{2^*}{2^*-1}} \, dx\right)^{\frac{2^*-1}{2^*}} \to 0$$

as $k \to +\infty$. Recall that we are working with a subsequence of the original sequence u_k. Hence, what we have really proved is that for every sequence $u_k \to u$ there is a subsequence $\{u_{k_j}\}_j$ such that $J'_G(u_{k_j}) \to J'_G(u)$ in $[H^1(\Omega)]'$; from this it is an elementary exercise to conclude that $J'_G(u_k) \to J'_G(u)$ in $[H^1(\Omega)]'$. By Proposition 1.3.8 we deduce that J is differentiable and $J'(u)v = \int_\Omega f(u)v \, dx$.

Example 1.3.21 If we drop the hypothesis that Ω is bounded we have to impose stronger conditions on f. So, let $\Omega \subseteq \mathbb{R}^N$, $N \geq 3$, be a general open set (with smooth boundary). Let $f : \mathbb{R} \to \mathbb{R}$ be a continuous function, and assume that there exist $a, b > 0$ such that

$$|f(t)| \leq a|t| + b|t|^{2^*-1} \tag{1.10}$$

for all $t \in \mathbb{R}$. Note that this assumption forces $f(0) = 0$.

Define, as above, $F(t) = \int_0^t f(s) \, ds$ and consider the same functional as in the preceding example, namely

$$J(u) = \int_\Omega F(u) \, dx.$$

Then J is differentiable on $H^1(\Omega)$ and

$$J'(u)v = \int_\Omega f(u)v \, dx \quad \text{for all } u, v \in H^1(\Omega).$$

As in the previous example, if we consider J on $H_0^1(\Omega)$, without any regularity assumption on $\partial\Omega$, then the same result holds. As before, let us first prove that J is Gâteaux differentiable. We have, for a.e. $x \in \Omega$,

$$\lim_{t \to 0} \frac{F(u(x) + tv(x)) - F(u(x))}{t} = f(u(x))v(x).$$

By the Lagrange Theorem, there exists a real number θ such that $|\theta| \leq |t|$ and

$$\left| \frac{F(u(x) + tv(x)) - F(u(x))}{t} \right| = \left| f(u + \theta v)v \right|$$

$$\leq C \left(|u + \theta v| + |u + \theta v|^{2^* - 1} \right) |v|$$

$$\leq C \left(|u||v| + |v|^2 + |u|^{2^* - 1}|v| + |v|^{2^*} \right) \in L^1(\Omega),$$

so we apply the dominated convergence theorem and we conclude as in the previous example that J is Gâteaux differentiable with $J_G'(u)v = \int_\Omega f(u)v\,dx$.

Again we assume that $\{u_k\}_k$ is a sequence in $H^1(\Omega)$ and $u_k \to u$ in $H^1(\Omega)$. Passing to a subsequence, we have that

- $u_k \to u$ in $L^{2^*}(\Omega)$ and in $L^2(\Omega)$;
- $u_k(x) \to u(x)$ a.e. in Ω;
- There exist $w_1 \in L^{2^*}(\Omega)$ and $w_2 \in L^2(\Omega)$ such that $|u_k(x)| \leq w_i(x)$ a.e. in Ω and for all $k \in \mathbb{N}$.

Let us now fix $\varepsilon > 0$ and take $R_\varepsilon > 0$ such that, setting $\Omega_\varepsilon = \{x \in \Omega \mid |x| > R_\varepsilon\}$, we have

$$|u|_{L^2(\Omega_\varepsilon)} + |u|_{L^{2^*}(\Omega_\varepsilon)}^{2^* - 1} + |w_1|_{L^{2^*}(\Omega_\varepsilon)}^{2^* - 1} + |w_2|_{L^2(\Omega_\varepsilon)} \leq \varepsilon.$$

Now,

$$\left| \left(J_G'(u_k) - J_G'(u) \right) v \right| \leq \int_\Omega |f(u_k) - f(u)||v|\,dx$$

$$= \int_{\Omega \cap B_{R_\varepsilon}} |f(u_k) - f(u)||v|\,dx + \int_{\Omega_\varepsilon} |f(u_k) - f(u)||v|\,dx;$$

on Ω_ε we have

$$\int_{\Omega_\varepsilon} |f(u_k) - f(u)||v|\,dx$$

$$\leq C \int_{\Omega_\varepsilon} \left(|u_k| + |u| + |u_k|^{2^* - 1} + |u|^{2^* - 1} \right) |v|\,dx$$

$$\leq C \left(\int_{\Omega_\varepsilon} |w_2||v|\,dx + \int_{\Omega_\varepsilon} |u||v|\,dx + \int_{\Omega_\varepsilon} |w_1|^{2^* - 1}|v|\,dx + \int_{\Omega_\varepsilon} |u|^{2^* - 1}|v|\,dx \right)$$

$$\leq C\|v\| \left(|w_2|_{L^2(\Omega_\varepsilon)} + |u|_{L^2(\Omega_\varepsilon)} + |w_1|_{L^{2^*}(\Omega_\varepsilon)}^{2^* - 1} + |u|_{L^{2^*}(\Omega_\varepsilon)}^{2^* - 1} + \right) \leq C\|v\|\varepsilon.$$

On the set $\Omega \cap B_{R_\varepsilon}$ we have

$$\int_{\Omega \cap B_{R_\varepsilon}} |f(u_k) - f(u)||v|\,dx \leq C \left(\int_{\Omega \cap B_{R_\varepsilon}} |f(u_k) - f(u)|^{\frac{2^*}{2^* - 1}}\,dx \right)^{\frac{2^* - 1}{2^*}} \|v\|,$$

and since from the assumption on f it readily follows that there is $C > 0$ such that $|f(t)| \leq C(1 + |t|^{2^* - 1})$ for all t, we then obtain as in the preceding example that

$$\int_{\Omega \cap B_{R_\varepsilon}} |f(u_k) - f(u)|^{\frac{2^*}{2^* - 1}}\,dx \to 0$$

as $k \to \infty$. Thus

$$\|J'_G(u_k) - J'_G(u)\| = \sup\left\{\left|(J'_G(u_k) - J'_G(u))v\right| \mid v \in H^1(\Omega), \|v\| = 1\right\}$$

$$\leq C\left(\int_{\Omega \cap B_{R_\varepsilon}} |f(u_k) - f(u)|^{\frac{2^*}{2^*-1}} dx\right)^{\frac{2^*-1}{2^*}} + C\varepsilon = o(1) + C\varepsilon$$

as $k \to \infty$. This means that

$$\limsup_{k \to +\infty} \|J'_G(u_k) - J'_G(u)\| \leq C\varepsilon.$$

Since this happens for all $\varepsilon > 0$, so we deduce

$$\limsup_{k \to +\infty} \|J'_G(u_k) - J'_G(u)\| = 0.$$

This argument holds for a subsequence of the original sequence $\{u_k\}_k$; we conclude the as in the previous case.

1.4 Weak Solutions and Critical Points

The variational approach to semilinear elliptic equations is based on the notion of *weak solution*.

For a general discussion on the importance of weakening the notion of solution in many areas of differential equations, we refer the reader to [18]. Here, we limit ourselves to the description of the ideas involved around this notion for the specific equations we deal with.

It is important that the reader bears in mind this fact: to each differential equation one can associate a notion of weak solution, or even more than one, depending on what one is interested in. Of course, there are *classes* of differential equations that share the same definition of weak solution, such as almost all the problems treated in this book.

It is customary to start with a simple linear problem that serves as an example and a guideline for more sophisticated cases.

Let Ω be a bounded open set in $\mathbb{R}^N$, and let $q, h \in C(\Omega)$.

Suppose we want to find a function $u : \overline{\Omega} \to \mathbb{R}$ such that

$$\begin{cases} -\Delta u + q(x)u = h(x) & \text{in } \Omega, \\ u = 0 & \text{on } \partial\Omega, \end{cases} \tag{1.11}$$

where Δ is the Laplace operator, defined in (1.4).

This problem is called the *homogeneous Dirichlet problem*. Generally speaking, the Dirichlet problem consists in coupling a differential equation with a boundary condition that specifies the values of the unknown function on the boundary of Ω; one says that the Dirichlet condition is homogeneous if the unknown is required to be zero on $\partial\Omega$. The homogeneous Dirichlet problem is the prototype of all boundary value problems, and this is why in this book we only consider this type of boundary conditions.

A *classical solution* of (1.11) is a function $u \in C^2(\overline{\Omega})$ that satisfies (1.11) for every $x \in \overline{\Omega}$. We now make the following experiment: we take $v \in C_0^1(\Omega)$, we multiply the equation in (1.11) by v and we integrate over Ω. By the Green's formula (1.5) (notice that the boundary integral vanishes, since v has compact support) we obtain that if u is a classical solution, then

$$\int_\Omega \nabla u \cdot \nabla v \, dx + \int_\Omega q(x) u v \, dx = \int_\Omega h(x) v \, dx, \quad \forall v \in C_0^1(\Omega). \qquad (1.12)$$

Now this formula makes sense even if u is not C^2; for example $u \in C^1$ suffices. On a closer inspection, one realizes that the regularity requirements on u *and* v can still be weakened very much. Indeed for the integrals to be finite it is enough that $u, v \in L^2(\Omega)$ and so do $\frac{\partial u}{\partial x_i}$ and $\frac{\partial v}{\partial x_i}$, for every i. Having observed this, it is even no longer necessary for q and h to be continuous: one can merely require that $q \in L^\infty(\Omega)$, and $h \in L^2(\Omega)$.

This motivates the following fundamental definition.

Definition 1.4.1 Let $q \in L^\infty(\Omega)$ and $h \in L^2(\Omega)$. A weak solution of problem (1.11) is a function $u \in H_0^1(\Omega)$ such that

$$\int_\Omega \nabla u \cdot \nabla v \, dx + \int_\Omega q(x) u v \, dx = \int_\Omega h(x) v \, dx, \quad \forall v \in H_0^1(\Omega). \qquad (1.13)$$

A few comments are in order. If one wants to *extend* the notion of solution, one would like, to say the least, that classical solutions be weak solutions as well. This is indeed true in the present case, as we now show. Let u be a classical solution of (1.11). In particular, $u \in C^2(\overline{\Omega})$, so that $u \in H^1(\Omega)$. Since u is continuous on $\overline{\Omega}$ and $u = 0$ on $\partial\Omega$, then $u \in H_0^1(\Omega)$, see the definition of $H_0^1(\Omega)$ in Sect. 1.2.1. Hence u is in the right function space. Next, we know that (1.12) holds, and since $C_0^1(\Omega)$ is dense in $H_0^1(\Omega)$, for any fixed $v \in H_0^1(\Omega)$ we can take a sequence $\{v_n\}_n \subset C_0^1(\Omega)$ such that $v_n \to v$ in the H_0^1 topology. Letting $n \to \infty$, we obtain that (1.12) holds for every $v \in H_0^1(\Omega)$. This is exactly (1.13), and hence u is a weak solution.

Notice that, apparently, in the above definition there is no specification of boundary conditions. Actually the homogeneous boundary condition is hidden, or better, englobed in the functional setting of the problem: all functions in $H_0^1(\Omega)$ satisfy $u = 0$ on $\partial\Omega$, in an appropriate sense (see again the definition of $H_0^1(\Omega)$ in Sect. 1.2.1). This is one of the useful features of the notion of weak solution.

Now suppose we have found a weak solution u of (1.11), so that u satisfies the integral relation (1.13). Is there any chance that u be a *classical* solution? Of course if we are interested in classical solutions we suppose from the beginning that q and h are continuous. A very important aspect in the definition of weak solution is the fact that the answer to this question only depends on the *regularity* of u. Indeed it is easy to prove that if u is a weak solution *and* u is in $C^2(\overline{\Omega})$, then u is a classical solution. To see this notice first of all that $u \in H_0^1(\Omega) \cap C^2(\overline{\Omega})$ implies $u = 0$ on $\partial\Omega$ in the classical sense. Next, one can take $v \in C_0^1(\Omega)$ in (1.13), obtaining that (1.12) holds. Then one can use the Green's formula as above, but in the opposite direction, to show that

$$\int_\Omega \big(-\Delta u + q(x)u - h(x)\big)v\,dx = 0, \quad \forall v \in C_0^1(\Omega). \tag{1.14}$$

Since $C_0^1(\Omega)$ is dense in $L^2(\Omega)$, we see that

$$\begin{cases} -\Delta u + q(x)u = h(x) & \text{a.e. in } \Omega, \\ u = 0 & \text{on } \partial\Omega, \end{cases} \tag{1.15}$$

but since u is C^2, the equation holds at every point. Hence u is a classical solution.

The preceding discussion highlights a further fundamental feature of the weak solution approach: existence is completely separated from regularity. One can concentrate on existence results (and this is what this book is all about) and only *later* be concerned about regularity of weak solutions. When everything works one finds a classical solution of the problem. The trick is that regularity theory for weak solutions of partial differential equations is a quite hard and technical issue. However there are now precise results that one can invoke in the majority of situations. We do not discuss these items here, and we refer the reader to [11, 18, 22, 24, 45] for precise regularity results.

We now close the circle by showing how weak solutions are related to critical points of functionals, still for problem (1.11).

Define a functional $J : H_0^1(\Omega) \to \mathbb{R}$ by

$$J(u) = \frac{1}{2}\int_\Omega |\nabla u|^2\,dx + \frac{1}{2}\int_\Omega q(x)u^2\,dx - \int_\Omega h(x)u\,dx. \tag{1.16}$$

This functional is often called the *energy functional* associated to problem (1.11). The term is borrowed from the applications, where J is likely to represent an energy of some sort. Notice that the first two terms are quadratic in u, while the third is linear. By the results of Sect. 1.3, and in particular of Example 1.3.17, the functional J is differentiable on $H_0^1(\Omega)$, and

$$J'(u)v = \int_\Omega \nabla u \cdot \nabla v\,dx + \int_\Omega q(x)uv\,dx - \int_\Omega h(x)v\,dx, \quad \forall v \in H_0^1(\Omega). \tag{1.17}$$

Now the connection between weak solutions and critical points is evident: comparing Definition 1.4.1 and (1.17), one sees that *u is a weak solution of problem* (1.11) *if and only if u is a critical point of the functional J.*

Thus we have here a concrete example of the Euler–Lagrange equation associated to the functional J, see Definition 1.3.10.

The correspondence between weak solutions and critical point of functionals outlined above is valid of course for more general nonlinear problems. The procedure to define the notion of weak solution for certain nonlinear equations follows exactly the same steps as in the linear case: one "tests" the equation with a smooth function vanishing on the boundary of Ω, integrates by parts by means of the Green's formula, reduces the regularity requirement on the functions, and tries to interpret the integral equality as the vanishing of the differential of a suitable functional.

We now give a fundamental example by means of the results obtained in Sect. 1.3.

Let $\Omega \subset \mathbb{R}^N$, with $N \geq 3$, be open and bounded. Assume that $q \in L^\infty(\Omega)$ and let $f : \mathbb{R} \to \mathbb{R}$ be a continuous function that satisfies the growth condition (1.9):

$$|f(t)| \leq a + b|t|^{2^*-1}$$

for all $t \in \mathbb{R}$ and for some constants a, b. Recall that $2^* = \frac{2N}{N-2}$. Consider the problem

$$\begin{cases} -\Delta u + q(x)u = f(u) & \text{in } \Omega, \\ u = 0 & \text{on } \partial\Omega. \end{cases} \tag{1.18}$$

The procedure described for the linear case leads to the following definition.

Definition 1.4.2 A weak solution of problem (1.18) is a function $u \in H_0^1(\Omega)$ such that

$$\int_\Omega \nabla u \cdot \nabla v \, dx + \int_\Omega q(x)uv \, dx = \int_\Omega f(u)v \, dx, \quad \forall v \in H_0^1(\Omega). \tag{1.19}$$

Now we construct the energy functional associated to this problem. Let

$$F(t) = \int_0^t f(s) \, ds$$

and consider the functional $J : H_0^1(\Omega) \to \mathbb{R}$ defined by

$$J(u) = \frac{1}{2} \int_\Omega |\nabla u|^2 \, dx + \frac{1}{2} \int_\Omega q(x)u^2 \, dx - \int_\Omega F(u) \, dx.$$

By the results of Sect. 1.3, J is differentiable on $H_0^1(\Omega)$, and

$$J'(u)v = \int_\Omega \nabla u \cdot \nabla v \, dx + \int_\Omega q(x)uv \, dx - \int_\Omega f(u)v \, dx, \quad \forall v \in H_0^1(\Omega).$$

Therefore weak solutions of (1.18) correspond to critical points of J.

Remark 1.4.3 The growth condition (1.9) is essential to have that the energy functional is well-defined on $H_0^1(\Omega)$. Indeed, if $f(t)$ grows faster that $|t|^{2^*-1}$, then $F(t)$ grows faster than $|t|^{2^*}$; since $H_0^1(\Omega)$ is *not* embedded in $L^p(\Omega)$ when $p > 2^*$, the integral of $F(u)$ in the functional J might diverge for some $u \in H_0^1(\Omega)$.

1.5 Convex Functionals

Let us begin with a heuristic "principle". The typical functionals $I(u)$ whose critical points give rise to weak solutions of differential equations are integral functionals that normally contain a term involving the gradient of u, in many cases the integral of some power of $|\nabla u|$. In a physical or mechanical context this term often represents an *energy* of some sort. This type of term is *bounded below* because it is nonnegative, but in general it is *not bounded above* (energies can be made arbitrarily high). Thus if a functional contains such a term, it may be perhaps *minimized*, but probably not maximized.

This is why the most "natural" critical points of differentiable integral functionals are often its *global minima*.

Remark 1.5.1 A (local) point of minimum of a differentiable functional is of course a critical point. Indeed, if $I : X \to \mathbb{R}$ is a differentiable functional on a Banach space X and

$$I(u) = \min_{v \in X} I(v),$$

then, for every fixed $v \in X$, we can consider the differentiable function $\phi : \mathbb{R} \to \mathbb{R}$ defined by $\phi(t) = I(u + tv)$. The function ϕ has a (local) minimum at $t = 0$ and is differentiable. Therefore

$$0 = \phi'(0) = I'(u)v.$$

This holds for every $v \in X$, and hence u is a critical point for I.

In the search for global minima a relevant concept is convexity.

Definition 1.5.2 A functional $I : X \to \mathbb{R}$ on a vector space X is called convex if for every $u, v \in X$ and every real $t \in [0, 1]$ there results

$$I(tu + (1 - t)v) \leq t I(u) + (1 - t) I(v).$$

The functional is called strictly convex if for every $u, v \in X$, $u \neq v$, and every real $t \in (0, 1)$ there results

$$I(tu + (1 - t)v) < t I(u) + (1 - t) I(v).$$

Finally, we say that I is (strictly) concave if $-I$ is (strictly) convex.

The main result for our purposes is the following classical theorem (see [11]).

Theorem 1.5.3 *Let $I : X \to \mathbb{R}$ be a continuous convex functional on a Banach space X. Then I is weakly lower semicontinuous. In particular, for every sequence $\{u_k\}_{k \in \mathbb{N}} \subset X$ converging weakly to $u \in X$, we have*

$$I(u) \leq \liminf_{k \to \infty} I(u_k).$$

Remark 1.5.4 On a Banach space X the norm is an example of a continuous convex functional. Therefore the preceding result can be applied, obtaining the following fact that we will use repeatedly in the sequel: if $u_k \rightharpoonup u$ in X, then

$$\|u\| \leq \liminf_{k \to \infty} \|u_k\|.$$

A continuous convex functional need not have a minimum, even if it is bounded below, and even in finite dimension: think for example of the function e^x. The problem is that minimizing sequences may "escape to infinity". The missing ingredient is given by the following notion.

Definition 1.5.5 A functional $I : X \to \mathbb{R}$ on a Banach space X is called coercive if, for every sequence $\{u_k\}_{k\in\mathbb{N}} \subset X$,

$$\|u_k\| \to +\infty \quad \text{implies} \quad I(u_k) \to +\infty.$$

A functional I is called anticoercive if $-I$ is coercive.

The following result is of fundamental importance.

Theorem 1.5.6 *Let X be a reflexive Banach space and let $I : X \to \mathbb{R}$ be a continuous, convex and coercive functional. Then I has a global minimum point.*

Proof Define $m = \inf_{u\in X} I(u)$. Let $\{u_k\}_{k\in\mathbb{N}} \subset X$ be a minimizing sequence. Coercivity implies that $\{u_k\}_{k\in\mathbb{N}}$ is bounded. Since X is reflexive, by the Banach–Alaoglu Theorem we can extract from $\{u_k\}_{k\in\mathbb{N}}$ a subsequence, still denoted u_k, such that u_k converges weakly to some $u \in X$. By Theorem 1.5.3 we then obtain

$$I(u) \le \liminf_{k\to\infty} I(u_k) = m.$$

Therefore $I(u) = m$ and u is a global minimum for I. $\qquad\qquad\qquad\square$

Remark 1.5.7 In the previous statement the convexity assumption is only used to deduce weak lower semicontinuity from continuity (via Theorem 1.5.3). A more general statement is thus the following version of the *Weierstrass Theorem*: let X be a reflexive Banach space and let $I : X \to \mathbb{R}$ be weakly lower semicontinuous and coercive. Then I has a global minimum point. This is the starting point of the so-called *direct methods of the Calculus of Variations*.

Convexity is also strongly related to uniqueness properties. We present two results in this direction.

Theorem 1.5.8 *Let $I : X \to \mathbb{R}$ be strictly convex. Then I has at most one minimum point in X.*

Proof Assume that I has two different global minima u_1 and u_2 in X. By strict convexity,

$$\min_{u\in X} I(u) \le I\left(\frac{u_1 + u_2}{2}\right) < \frac{1}{2}I(u_1) + \frac{1}{2}I(u_2) = \frac{1}{2}\min_{u\in X} I(u) + \frac{1}{2}\min_{u\in X} I(u)$$
$$= \min_{u\in X} I(u),$$

a contradiction. $\qquad\qquad\qquad\qquad\qquad\qquad\qquad\qquad\qquad\qquad\qquad\qquad\square$

Theorem 1.5.9 *Let* $I : X \to \mathbb{R}$ *be strictly convex and differentiable. Then* I *has at most one critical point in* X.

Proof The claim is obvious if $X = \mathbb{R}$, that is, if I is a differentiable real function of one real variable. Indeed, in this case strict convexity implies that I' is strictly increasing, and hence there is at most one $t \in \mathbb{R}$ such that $I'(t) = 0$. In the general case, assume that u is a critical point for I, fix any $v \in X$ and define $\gamma : \mathbb{R} \to \mathbb{R}$ by $\gamma(t) = I(u + tv)$. The function γ is differentiable, and it is easy to see that it is strictly convex and that $\gamma'(0) = 0$. Therefore $\gamma'(t) \neq 0$ for all $t \neq 0$, which means $I'(u + tv)v \neq 0$, and hence $I'(u + tv) \neq 0$ for all $t \neq 0$. As this holds for every $v \in X$, the result follows. $\qquad\square$

We conclude with a useful criterion to detect convexity.

Proposition 1.5.10 *Let* X *be a Banach space and let* $I : X \to \mathbb{R}$ *be a differentiable functional. Assume that for all* $u, v \in X$,

$$(I'(u) - I'(v))(u - v) \geq 0.$$

Then I *is convex. If the strict inequality holds when* $u \neq v$, *then* I *is strictly convex.*

Proof Fix $u, v \in X$ and define a function $\psi : \mathbb{R} \to \mathbb{R}$ as

$$\psi(t) = I(u + t(v - u)).$$

The function ψ is differentiable and $\psi'(t) = I'(u + t(v - u))(v - u)$. Fix now $s < t$. We have

$$\psi'(t) - \psi'(s) = \left[I'(u + t(v - u)) - I'(u + s(v - u)) \right](v - u)$$

$$= \frac{1}{t - s}\left[I'(u + t(v - u)) - I'(u + s(v - u)) \right]$$

$$\times \left[(u + t(v - u)) - (u + s(v - u)) \right] \geq 0.$$

Hence ψ' is a nondecreasing function, so ψ is convex, and in particular

$$\psi(t) = \psi(1t + 0(1 - t)) \leq t\psi(1) + (1 - t)\psi(0),$$

which means $I(tv + (1 - t)u) \leq tI(v) + (1 - t)I(u)$.

If in the assumption the strict inequality holds, then we get that ψ' is strictly increasing, and hence ψ is strictly convex, implying the same for I. $\qquad\square$

1.6 A Few Examples

Let us now apply the abstract results of the previous sections to some concrete differential problems.

Theorem 1.6.1 *Let $\Omega \subset \mathbb{R}^N$ be a bounded open set and let $A(x)$ be a symmetric $N \times N$ matrix-valued function with entries in $L^\infty(\Omega)$. Assume that there exists $\lambda > 0$ such that $(A(x)y)y \geq \lambda |y|^2$ for a.e. $x \in \Omega$ and for all $y \in \mathbb{R}^N$. Let $q \in L^\infty(\Omega)$ satisfy $q(x) \geq 0$ a.e. in Ω. Then for every $h \in L^2(\Omega)$ the problem*

$$\begin{cases} -\operatorname{div}\big(A(x)\nabla u\big) + q(x)u = h(x) & \text{in } \Omega, \\ u = 0 & \text{on } \partial\Omega \end{cases} \tag{1.20}$$

has a unique (weak) solution.

Remark 1.6.2 The condition $(A(x)y)y \geq \lambda|y|^2$ for a.e. $x \in \Omega$ and for all $y \in \mathbb{R}^N$ states that the matrix $A(x)$ is (uniformly) positive definite on Ω. This, by Definition 1.2.3, says that the differential operator $-\operatorname{div}(A(x)\nabla u)$ is uniformly elliptic on Ω.

Proof Consider the functional $J : H_0^1(\Omega) \to \mathbb{R}$ defined as

$$J(u) = \frac{1}{2}\int_\Omega (A(x)\nabla u)\cdot\nabla u\,dx + \frac{1}{2}\int_\Omega q(x)u^2\,dx - \int_\Omega hu\,dx.$$

We have seen that J is differentiable on $H_0^1(\Omega)$ and that

$$J'(u)v = \int_\Omega (A(x)\nabla u)\cdot\nabla v\,dx + \int_\Omega q(x)uv\,dx - \int_\Omega hv\,dx.$$

Thus a critical point of J is a weak solution of problem (1.20).

Since for $u \neq v$,

$$\begin{aligned}(J'(u) - J'(v))(u - v) &= \int_\Omega \big(A(x)(\nabla u - \nabla v)\big)\cdot(\nabla u - \nabla v)\,dx \\ &\quad + \int_\Omega q(x)(u - v)^2\,dx \\ &\geq \lambda \int_\Omega |\nabla(u - v)|^2\,dx > 0,\end{aligned}$$

J is strictly convex in view of Proposition 1.5.10.

As Ω is bounded, thanks to the Poincaré and Sobolev inequalities, we have

$$J(u) \geq \frac{\lambda}{2}\int_\Omega |\nabla u|^2\,dx - |h|_2|u|_2 \geq C\|u\|^2 - C\|u\|,$$

and J is also coercive. By Theorems 1.5.6 and 1.5.8, J has a global minimum which is its only critical point. Therefore Problem (1.20) has a unique solution. □

In this result the sign condition on q is not *necessary* to obtain the desired properties. Indeed, the same conclusion holds provided q is "not too negative", for instance in the following sense.

Let S be the best Sobolev constant for the embedding of $H_0^1(\Omega)$ into $L^{2^*}(\Omega)$, that is, the largest positive constant S such that

$$S|u|_{2^*}^2 \leq \|u\|^2 \quad \text{for every } u \in H_0^1(\Omega). \tag{1.21}$$

Assume that $q \in L^{N/2}(\Omega)$ satisfies

$$|q|_{N/2} < \lambda S.$$

Then the conclusion of the previous theorem holds. Indeed we note that by the Hölder inequality, for every $w \in H_0^1(\Omega)$,

$$\left| \int_\Omega q(x) w^2 \, dx \right| \le \left(\int_\Omega |q|^{\frac{N}{2}} \, dx \right)^{\frac{2}{N}} \left(\int_\Omega |w|^{\frac{2N}{N-2}} \, dx \right)^{\frac{N-2}{N}}$$

$$= |q|_{N/2} |w|_{2*}^2 \le \frac{1}{S} |q|_{N/2} \|w\|^2.$$

Then we have, with the same argument as above,

$$(J'(u) - J'(v)) (u - v) \ge \left(\lambda - \frac{1}{S} |q|_{N/2} \right) \|u - v\|^2 > 0,$$

which proves (strict) convexity, and

$$J(u) \ge \frac{\lambda}{2} \int_\Omega |\nabla u|^2 \, dx - \frac{1}{2S} |q|_{N/2} \|u\|^2 - |h|_2 |u|_2$$

$$\ge \frac{1}{2} \left(\lambda - \frac{1}{S} |q|_{N/2} \right) \|u\|^2 - C\|u\|,$$

which proves coercivity. Since under the weakened condition on q the functional J is still differentiable, as it is easy to check, we again obtain the existence of a solution to Problem (1.20).

A particular but significant case of the preceding theorem is worth to be stated separately.

Corollary 1.6.3 *Let $\Omega \subset \mathbb{R}^N$ be a bounded open set. For every $h \in L^2(\Omega)$ the problem*

$$\begin{cases} -\Delta u = h(x) & \text{in } \Omega, \\ u = 0 & \text{on } \partial\Omega. \end{cases} \tag{1.22}$$

has a unique (weak) solution.

Similarly, with the same arguments of Theorem 1.6.1, one can prove a version for a problem on $\mathbb{R}^N$.

Corollary 1.6.4 *For every $h \in L^2(\mathbb{R}^N)$ the problem*

$$\begin{cases} -\Delta u + u = h(x) & \text{in } \mathbb{R}^N, \\ u \in H^1(\mathbb{R}^N) \end{cases} \tag{1.23}$$

has a unique (weak) solution.

Remark 1.6.5 The solutions mentioned in these two corollaries are of course obtained as the unique global minima of the functionals $J : H_0^1(\Omega) \to \mathbb{R}$ defined by

$$J(u) = \frac{1}{2} \int_\Omega |\nabla u|^2 \, dx - \int_\Omega hu \, dx = \frac{1}{2} \|u\|^2 - \int_\Omega hu \, dx$$

and $I : H^1(\mathbb{R}^N) \to \mathbb{R}$ defined by

$$I(u) = \frac{1}{2}\int_{\mathbb{R}^N} |\nabla u|^2 \, dx + \frac{1}{2}\int_{\mathbb{R}^N} u^2 \, dx - \int_{\mathbb{R}^N} hu \, dx = \frac{1}{2}\|u\|^2 - \int_{\mathbb{R}^N} hu \, dx.$$

This procedure is called the *Dirichlet principle*.

We now give a first example of a *nonlinear* problem which can easily solved with the same techniques.

Theorem 1.6.6 *Let* $\Omega \subset \mathbb{R}^N$, *with* $N \geq 3$, *be open and bounded. Assume that* $q \in L^\infty(\Omega)$ *satisfies* $q \geq 0$ *a.e. in* Ω, *and let* $f : \mathbb{R} \to \mathbb{R}$ *be a continuous function such that* (1.9) *holds. If*

$$f(t)t \leq 0 \quad and \quad (f(t) - f(s))(t - s) \leq 0 \quad \forall t, s \in \mathbb{R}, \tag{1.24}$$

then, for every $h \in L^2(\Omega)$, *the problem*

$$\begin{cases} -\Delta u + q(x)u = f(u) + h(x) & in \ \Omega, \\ u = 0 & on \ \partial\Omega \end{cases} \tag{1.25}$$

has exactly one solution.

Proof Let $F(t) = \int_0^t f(s)\,ds$ and define a differentiable functional $J : H_0^1(\Omega) \to \mathbb{R}$ by

$$J(u) = \frac{1}{2}\int_\Omega |\nabla u|^2 \, dx + \frac{1}{2}\int_\Omega q(x)u^2 \, dx - \int_\Omega F(u)\,dx - \int_\Omega hu\,dx.$$

Critical points of J are weak solutions of (1.25). We have, for all $u, v \in H_0^1(\Omega)$,

$$(J'(u) - J'(v))(u - v) = \|u - v\|^2 - \int_\Omega (f(u) - f(v))(u - v)\,dx \geq \|u - v\|^2,$$

so that J is strictly convex. Furthermore it is easy to see that $F(t) \leq 0$ for all t, and hence

$$J(u) \geq \frac{1}{2}\|u\|^2 - \int_\Omega F(u)\,dx - |h|_2|u|_2 \geq \frac{1}{2}\|u\|^2 - C\|u\|,$$

by the Poincaré and Sobolev inequalities. This shows that J is also coercive. As the functional is coercive, convex and continuous, it has a global minimum point, which is a critical point. Since J is strictly convex, this is the only critical point, by Proposition 1.5.9. $\qquad\square$

As obvious examples of functions f satisfying the assumption of the previous theorem, one can take any $f(t) = -|t|^{p-2}t$ with $p \in (1, 2^*]$, or $f(t) = -\arctan t$.

1.7 Some Spectral Properties of Elliptic Operators

We now recall some well-known spectral properties of elliptic operators; for the sake of simplicity and in view of later applications we state the results for an operator of the form $-\Delta + q(x)$, the prototype of all (uniformly) elliptic operators. The

results can be extended with minor modifications to cases where e.g. the Laplacian
is replaced by the operator $-\operatorname{div}(A(x)\nabla)$, under suitable natural assumptions on the
matrix A.

Let Ω be a *bounded* open subset of $\mathbb{R}^N$, and let $q \in L^\infty(\Omega)$.

Definition 1.7.1 We say that $\lambda \in \mathbb{R}$ is an eigenvalue of the operator $-\Delta + q(x)$
under Dirichlet boundary conditions if there exists $\varphi \in H_0^1(\Omega) \setminus \{0\}$ such that

$$\begin{cases} -\Delta\varphi + q(x)\varphi = \lambda\varphi & \text{in } \Omega, \\ \varphi = 0 & \text{on } \partial\Omega. \end{cases} \tag{1.26}$$

The function φ is called an eigenfunction associated to λ.

Remark 1.7.2 Problem (1.26) is to be interpreted in the weak sense, namely φ solves
it provided

$$\int_\Omega \nabla\varphi \cdot \nabla v \, dx + \int_\Omega q(x)\varphi v \, dx = \lambda \int_\Omega \varphi v \, dx \quad \text{for all } v \in H_0^1(\Omega).$$

If $q \equiv 0$, the corresponding numbers λ are simply called the eigenvalues of the
Laplacian (often even omitting "under Dirichlet boundary conditions").

The next theorem collects the basic properties of eigenvalues and eigenfunctions;
for its proof, see [18]. Since we present the theory for the Dirichlet problem only,
when we talk about eigenvalues we always mean that Dirichlet boundary conditions
are imposed.

Theorem 1.7.3 *Let $\Omega \subset \mathbb{R}^N$ be bounded and open and let $q \in L^\infty(\Omega)$. Then there
exist sequences $\{\lambda_k\}_k \subset \mathbb{R}$ and $\{\varphi_k\}_k \subset H_0^1(\Omega)$ such that*

1. *each λ_k is an eigenvalue of $-\Delta + q(x)$ and each φ_k is an eigenfunction corresponding to λ_k;*
2. *$\lambda_k \to +\infty$ as $k \to \infty$;*
3. *$\{\varphi_k\}_k$ is an orthonormal basis of $L^2(\Omega)$;*
4. *for every k, $\varphi_k \in L^\infty(\Omega)$ and $\varphi_k(x) \neq 0$ a.e. in Ω.*

Remark 1.7.4 It is customary to list the eigenvalues in an infinite increasing sequence $\lambda_1 \leq \lambda_2 \leq \cdots$; thus eigenvalues are repeated according to their (finite) multiplicity. Two eigenfunctions φ_i and φ_j are *always* different for $i \neq j$ (they are orthogonal in L^2), even if they correspond to the *same* eigenvalue.

Notation For a fixed operator on a fixed domain, eigenvalues are denoted by λ_k.
If one wants to stress the dependence of λ_k on the operator or on the domain one
writes $\lambda_k(-\Delta + q)$ or $\lambda_k(\Omega)$.

Remark 1.7.5 When $\lambda_1(-\Delta + q) > 0$ (this happens certainly if $q(x) \geq 0$ a.e. in Ω,
by the characterization of λ_1 in Theorem 1.7.6 below), it is easy to see that the
quantity

$$(u|v) = \int_\Omega \nabla u \cdot \nabla v \, dx + \int_\Omega q(x)uv \, dx$$

defines a scalar product on $H_0^1(\Omega)$ that induces a norm $\|\cdot\|$ equivalent to the usual one. One checks immediately that the set $\{\frac{\varphi_k}{\sqrt{\lambda_k}}\}_k$ is an orthonormal basis of $H_0^1(\Omega)$ with respect to the scalar product defined above.

For future reference we notice that we can write

$$u = \sum_{k=1}^{\infty} \alpha_k \varphi_k \quad \text{in } L^2(\Omega) \quad \text{and} \quad |u|_2^2 = \sum_{k=1}^{\infty} \alpha_k^2,$$

with $\alpha_k = \int_{\Omega} u \varphi_k \, dx$, or

$$u = \sum_{k=1}^{\infty} \beta_k \frac{\varphi_k}{\sqrt{\lambda_k}} \quad \text{in } H_0^1(\Omega) \quad \text{and} \quad \|u\|^2 = \sum_{k=1}^{\infty} \beta_k^2,$$

with $\beta_k = (u|\frac{\varphi_k}{\sqrt{\lambda_k}})$. The numbers α_k and β_k are the *Fourier coefficients* of u in $L^2(\Omega)$ and in $H_0^1(\Omega)$ respectively. Of course, $\beta_k = \sqrt{\lambda_k} \alpha_k$.

Eigenvalues and eigenfunctions play a very relevant role in nonlinear elliptic problems. For this reason we add some more information.

The most "important" eigenvalue is the smallest, λ_1, also called *the first eigenvalue* or *the principal eigenvalue*. It enjoys a number of special features, some of which we list in the next result.

Theorem 1.7.6 (Variational characterization of the first eigenvalue) *Let Ω be an open and bounded subset of $\mathbb{R}^N$ and let $q \in L^{\infty}(\Omega)$. Define a functional $Q : H_0^1(\Omega) \setminus \{0\} \to \Omega$ as*

$$Q(u) = \frac{\int_{\Omega} |\nabla u|^2 \, dx + \int_{\Omega} q(x) u^2 \, dx}{\int_{\Omega} u^2 \, dx}.$$

This functional is called the Rayleigh quotient. Then

1. *$\min_{u \in H_0^1(\Omega) \setminus \{0\}} Q(u) = \lambda_1$;*
2. *$Q(u) = \lambda_1$ if and only if u is a (weak) solution of*

$$\begin{cases} -\Delta u + q(x)u = \lambda_1 u & \text{in } \Omega, \\ u = 0 & \text{on } \partial\Omega; \end{cases} \tag{1.27}$$

3. *every non identically zero solution of (1.27) has constant sign in Ω (in particular, is a.e. different from zero in Ω);*
4. *the set of solutions of (1.27) has dimension one; one says that λ_1 is simple.*

The fact that the first eigenvalue is simple is normally expressed by writing the sequence of eigenvalues as $\lambda_1 < \lambda_2 \leq \lambda_3 \leq \cdots$, with the strict inequality between the first two terms. Since the eigenfunctions associated to λ_1 are all of the form

$\theta\varphi_1$, for $\theta \in \mathbb{R}$ and for some $\varphi_1 \in H_0^1(\Omega)$, it is customary to say that φ_1 is *the* first eigenfunction, in the sense that there exists a unique $\varphi_1 \in H_0^1(\Omega)$ such that

$$\begin{cases} -\Delta\varphi_1 + q(x)\varphi_1 = \lambda_1\varphi_1 & \text{in } \Omega, \\ \varphi_1 = 0 & \text{on } \partial\Omega, \\ \varphi_1 > 0 & \text{in } \Omega, \\ \varphi_1 & \text{normalized.} \end{cases} \tag{1.28}$$

The normalization can be expressed by $\|\varphi_1\| = 1$, or $|\varphi_1|_2 = 1$ or by other similar statements, depending on the convenience of the moment.

Remark 1.7.7 All eigenfunctions but φ_1 change sign. Indeed, since they form an orthonormal basis of $L^2(\Omega)$, they satisfy in particular $\int_\Omega \varphi_1\varphi_k \, dx = 0$ for $k \neq 1$. As φ_1 is positive, φ_k must change sign.

The role of eigenvalues is of fundamental importance for the solvability of linear problems. We give the main result, that can be seen as a direct application of the Fredholm alternative in Functional Analysis (see [11, 18]).

Theorem 1.7.8 *Let Ω be a bounded open subset of $\mathbb{R}^N$. Assume that $q \in L^\infty(\Omega)$ and that $h \in L^2(\Omega)$. Consider the problem*

$$\begin{cases} -\Delta u + q(x)u = \lambda u + h & \text{in } \Omega, \\ u = 0 & \text{on } \partial\Omega. \end{cases} \tag{1.29}$$

Then

1. *Problem (1.29) has a unique (weak) solution for every $h \in L^2(\Omega)$ if and only if λ is not an eigenvalue of $-\Delta + q(x)$.*
2. *If λ is an eigenvalue, then problem (1.29) has a solution if and only if h is orthogonal in $L^2(\Omega)$ to $\ker(-\Delta + q(x) - \lambda)$.*

In this case, the general solution of (1.29) is of the form $u = u_0 + \varphi$, where u_0 is a fixed solution of (1.29) and φ is any function in $\ker(-\Delta + q(x) - \lambda)$.

As an application of the spectral properties discussed so far, we state a very useful property that allows one to prove easily that solutions of certain equations are nonnegative. This is quite important in many problems where by physical reasons, for example, one is only interested in positive solutions.

We are talking about a version of the *maximum principle*, a fundamental tool in the study of differential equations. For a complete discussion of this principle, we refer the reader to [18] or [41].

Proposition 1.7.9 *Let $\Omega \subset \mathbb{R}^N$ be open and bounded, and let $q \in L^\infty(\Omega)$. Assume that $u \in H_0^1(\Omega)$ satisfies*

$$-\Delta u + q(x)u \geq 0 \quad \text{in } \Omega$$

in the weak sense, that is,

$$\int_\Omega \nabla u \cdot \nabla \varphi \, dx + \int_\Omega q(x)u\varphi \, dx \geq 0 \quad \forall \varphi \in H_0^1(\Omega), \varphi(x) \geq 0 \text{ a.e. in } \Omega. \tag{1.30}$$

If $\lambda_1(-\Delta + q) > 0$, then $u(x) \geq 0$ almost everywhere in Ω.

Example 1.7.10 A typical use of the preceding proposition goes as follows. Suppose that the function $f : \Omega \times \mathbb{R} \to \mathbb{R}$ is continuous and satisfies $f(x, t) \geq 0$ for $t \geq 0$ and $f(x, 0) = 0$ (for every $x \in \Omega$). We are interested in nonnegative solutions of

$$\begin{cases} -\Delta u + q(x)u = f(x, u) & \text{in } \Omega, \\ u = 0 & \text{on } \partial\Omega. \end{cases} \tag{1.31}$$

We define

$$g(x, t) = \begin{cases} f(x, t) & \text{if } t \geq 0, \\ 0 & \text{if } t < 0. \end{cases}$$

Of course g is continuous. Then we solve (if possible) Problem (1.31) with f replaced by g. We obtain in this way a function $u \in H_0^1(\Omega)$ that satisfies

$$-\Delta u + q(x)u = g(x, u) \geq 0$$

at least in the weak sense. If $\lambda_1(-\Delta + q) > 0$, then by Proposition 1.7.9 we obtain that $u(x) \geq 0$ a.e. in Ω. But then, for almost every x in Ω, $g(x, u(x)) = f(x, u(x))$, so that actually u (is nonnegative) and solves (1.31).

Remark 1.7.11 Under regularity conditions, it is possible to prove that nonnegative solutions are actually everywhere positive in Ω. This is the content of the *strong maximum principle*, which is beyond the aims of this book. We refer the interested reader to [18, 41].

1.8 Exercises

1. Let $\Omega \subset \mathbb{R}^N$ be open and bounded. Consider the problem

$$\begin{cases} -\Delta u = 1 & \text{in } \Omega, \\ u = 0 & \text{on } \partial\Omega. \end{cases} \tag{1.32}$$

   (a) Define the notion of weak solution for this problem, and find a functional on $H_0^1(\Omega)$ whose critical points correspond to weak solutions of (1.32).
   (b) Show that (1.32) has exactly one solution.
   (c) Find this solution when Ω is a ball of radius R centered at zero. (*Hint:* write the equation in polar coordinates and look for a radial solution.)

2. Let $\Omega \subset \mathbb{R}^N$, with $N \geq 3$ be open and bounded. Take $q \in L^\infty(\Omega)$ and let $A(x)$ be an $N \times N$ matrix with continuous entries (on $\overline{\Omega}$).

   For $p \in (2, 2^*]$, consider the problem

$$\begin{cases} -\operatorname{div}(A(x)\nabla u) + q(x)u = |u|^{p-2}u & \text{in } \Omega, \\ u = 0 & \text{on } \partial\Omega. \end{cases} \tag{1.33}$$

Define the notion of weak solution for this problem, and find a functional on
$H_0^1(\Omega)$ whose critical points correspond to weak solutions of (1.33).

3. Let $f : \mathbb{R} \to \mathbb{R}$ be a continuous function satisfying (1.10) and (1.24). Assume
that $a \in L^\infty(\mathbb{R}^N)$ is such that

$$a(x) \geq c > 0$$

for a.e. $x \in \mathbb{R}^n$. Prove that for every $h \in L^2(\mathbb{R}^N)$, the problem

$$\begin{cases} -\Delta u + u = a(x) f(u) + h(x), \\ u \in H^1(\mathbb{R}^N) \end{cases}$$

has exactly one solution.

4. Let Ω be an open subset of $\mathbb{R}^N$, $N \geq 3$, and take a function $q \in L^{N/2}(\Omega)$. Prove
that the function $a : H_0^1(\Omega) \times H_0^1(\Omega) \to \mathbb{R}$ defined by

$$a(u, v) = \int_\Omega quv\, dx$$

is bilinear and continuous. Deduce that the functional $J : H_0^1(\Omega) \to \mathbb{R}$ defined
by

$$J(u) = \int_\Omega q\, u^2\, dx$$

is differentiable and compute J'.

5. (Best constant in the Poincaré inequality.) Let $\Omega \subset \mathbb{R}^N$ be open and bounded.
Prove that

$$\int_\Omega u^2\, dx \leq \frac{1}{\lambda_1} \int_\Omega |\nabla u|^2\, dx \quad \text{for all } u \in H_0^1(\Omega)$$

and the constant $1/\lambda_1$ is the best (i.e., smallest) possible.

6. (Dependence of λ_1 on the domain.) For every $\Omega \subset \mathbb{R}^N$ open and bounded, denote by $\lambda_1(\Omega)$ the first eigenvalue of the Laplacian on $H_0^1(\Omega)$. Prove that $\lambda_1(\Omega)$
is strictly decreasing with respect to inclusion, in the sense that

$$\Omega \subset \Omega' \quad \text{implies} \quad \lambda_1(\Omega) > \lambda_1(\Omega').$$

7. (Dependence of λ_1 on q.) Let Ω be open and bounded. Prove that
(a) $\lambda_1(-\Delta + q(x))$ is a nondecreasing function of q, in the sense that

$$p, q \in L^\infty(\Omega) \text{ and } p(x) \leq q(x) \text{ a.e.} \quad \text{imply} \quad \lambda_1(-\Delta + p) \leq \lambda_1(-\Delta + q);$$

(b) λ_1 is a continuous function of q, in the sense that

$$|\lambda_1(-\Delta + q) - \lambda_1(-\Delta + p)| \leq \|q - p\|_{L^\infty(\Omega)}.$$

8. Let Ω be open and bounded. For $h \in L^2(\Omega)$ and $\lambda \in \mathbb{R}$, define $I : H_0^1(\Omega) \to \mathbb{R}$
by

$$I(u) = \frac{1}{2} \int_\Omega |\nabla u|^2\, dx - \frac{\lambda}{2} \int_\Omega u^2\, dx - \int_\Omega hu\, dx.$$

Prove that I is coercive on $H_0^1(\Omega)$ if and only if $\lambda < \lambda_1$.

9. More generally, assume that Ω and h are in the previous exercise, let $q \in L^\infty(\Omega)$, and define $J : H_0^1(\Omega) \to \mathbb{R}$ by

$$J(u) = \frac{1}{2} \int_\Omega |\nabla u|^2 \, dx + \frac{1}{2} \int_\Omega q(x) u^2 \, dx - \int_\Omega h u \, dx.$$

(a) Prove that if $\lambda_1(-\Delta + q(x)) > 0$, then the expression

$$(u, v) = \int_\Omega \nabla u \nabla v \, dx + \int_\Omega q(x) u v \, dx$$

defines on $H_0^1(\Omega)$ a scalar product that induces a norm equivalent to the standard one.

(b) Prove that J is coercive on $H_0^1(\Omega)$ if and only if $\lambda_1(-\Delta + q(x)) > 0$.

(c) Deduce the result of the preceding exercise from this.

10. Prove Proposition 1.7.9. (*Hint*: test (1.30) with u^- and use the variational characterization of λ_1.)

1.9 Bibliographical Notes

- Section 1.1: A very gentle introduction to the Calculus of Variations is the book by Hildebrandt and Tromba [25]. For a history of the early Calculus of Variations, see Goldstine [23].

- Section 1.2: Proofs, details, and general theory can be found in almost any book in Functional Analysis. A good reference is Brezis [11], which is PDE-oriented. Topics more strictly related to differential equations can also be found in Evans [18].

- Section 1.3: The reader who wishes to see a more systematic treatment of differential calculus in normed space is referred to Ambrosetti and Prodi [4].

- Section 1.4: For a more complete discussion on weak solutions, and particularly for the extension of the definitions in this section to nonhomogeneous Dirichlet problems and Neumann problems, see Chap. IX of [11].

- Section 1.5: The development of the results and techniques outlined in this section, namely minimization of functionals under coercivity, convexity or semi-continuity properties is the *Calculus of Variations*, and particularly the so-called *direct methods* in the Calculus of Variations. This is a very wide topic that has reached a high degree of complexity, matched by outstanding success in solving huge classes of problems taken from everywhere in science. The readers interested in these methods can consult Giusti [21], especially the first chapter.

- Section 1.7: The missing proofs in this section can be found in [11, 18].

Chapter 2
Minimization Techniques: Compact Problems

Throughout this chapter we show how techniques based on minimization arguments can be used to establish existence results for various types of problems.

Our aim is not to describe the most general results, but to give a series of examples, and to show how simple techniques can be refined to treat more complex cases.

2.1 Coercive Problems

We begin with the following problem, that will provide our guideline through the whole chapter. We want to find a (weak) solution to

$$\begin{cases} -\Delta u + q(x)u = f(u) + h(x) & \text{in } \Omega, \\ u = 0 & \text{on } \partial\Omega. \end{cases} \tag{2.1}$$

In this section, the general framework is specified by the assumptions

$(\mathbf{h_1})$ $\Omega \subset \mathbb{R}^N$ is bounded and open, $q \in L^\infty(\Omega)$ and $q(x) \geq 0$ a.e. in Ω.
$(\mathbf{h_2})$ $h \in L^2(\Omega)$.

We equip $H_0^1(\Omega)$ with the scalar product

$$(u|v) = \int_\Omega \nabla u \cdot \nabla v \, dx + \int_\Omega q(x) uv \, dx, \tag{2.2}$$

and we denote by $\| \cdot \|$ the induced norm, equivalent to the standard one.

Remark 2.1.1 In assumption $(\mathbf{h_1})$, the requirement $q(x) \geq 0$ a.e. is used only to obtain $\lambda_1(-\Delta + q(x)) > 0$, which guarantees that (2.2) is indeed a scalar product and that the induced norm is equivalent to the standard norm of $H_0^1(\Omega)$, see Remark 1.7.5 and Exercise 9 in Chap. 1. Therefore in all the results of this chapter, and similarly in all the subsequent chapters, the assumption $q \geq 0$ could be replaced be the "abstract" condition

$$\lambda_1(-\Delta + q(x)) > 0, \tag{2.3}$$

M. Badiale, E. Serra, *Semilinear Elliptic Equations for Beginners*, Universitext,
DOI 10.1007/978-0-85729-227-8_2, © Springer-Verlag London Limited 2011

and everything would work perfectly well with no changes in the proofs. The point is, precisely, that (2.3) is abstract, and nobody knows for which general q's it is satisfied. We prefer, in this book, to assume an explicit sign condition on q, rather than an indirect one on λ_1. The reader should however keep in mind this clarification.

We begin by assuming the following hypothesis on the nonlinearity f.

(**h$_3$**) $f : \mathbb{R} \to \mathbb{R}$ is continuous and bounded.

Setting $F(t) = \int_0^t f(s)\,ds$, the computations carried out in Example 1.3.20 show that the functional $I : H_0^1(\Omega) \to \mathbb{R}$ defined by

$$I(u) = \frac{1}{2}\int_\Omega |\nabla u|^2\,dx + \frac{1}{2}\int_\Omega q(x)u^2\,dx - \int_\Omega F(u)\,dx - \int_\Omega hu\,dx$$

$$= \frac{1}{2}\|u\|^2 - \int_\Omega F(u)\,dx - \int_\Omega hu\,dx$$

is differentiable on $H_0^1(\Omega)$. Its critical points are the weak solutions of (2.1).

Note that unless F is concave, which we do not assume, the functional I needs not be convex.

Theorem 2.1.2 *Under the assumptions* (**h$_1$**)–(**h$_3$**), *Problem* (2.1) *admits at least one solution.*

Remark 2.1.3 The leading idea of the proof is that since f is bounded, the term $\int_\Omega F(u)\,dx$ should grow *at most linearly* with respect to $\|u\|$, as well as the last term. If this is true, the functional I can be seen as an "at most linear" perturbation of the quadratic term $\|u\|^2$. This suggests the existence of a global minimum. Let us see how all this really works.

Proof We break it into two steps. We make repeated use of Hölder and Sobolev inequalities.

Step 1. *The functional I is coercive.* Note first that since f is bounded, then

$$|F(t)| \le M|t|$$

for some $M > 0$ and all $t \in \mathbb{R}$. Hence

$$\left|\int_\Omega F(u)\,dx\right| \le M\int_\Omega |u|\,dx \le C\|u\|,$$

where the last inequality comes from the continuity of the embedding of $H_0^1(\Omega)$ into $L^1(\Omega)$. This confirms the idea of the linear growth as in the preceding remark. Thus

$$I(u) = \frac{1}{2}\|u\|^2 - \int_\Omega F(u)\,dx - \int_\Omega hu\,dx \ge \frac{1}{2}\|u\|^2 - C\|u\| - |h|_2|u|_2$$

$$\ge \frac{1}{2}\|u\|^2 - C\|u\|,$$

which shows that I is coercive.

Step 2. The infimum of I is attained. Set

$$m = \inf_{u \in H_0^1(\Omega)} I(u).$$

Step 1 shows that $m > -\infty$, although one does not really need this: it will follow automatically from the fact that it is attained.

Let $\{u_k\}_k \subset H_0^1(\Omega)$ be a minimizing sequence for I; from Step 1 we immediately see that $\{u_k\}_k$ is bounded in $H_0^1(\Omega)$, and therefore we can assume that there is a subsequence, still denoted u_k, such that

- $u_k \rightharpoonup u$ in $H_0^1(\Omega)$;
- $u_k \to u$ in $L^2(\Omega)$;
- $u_k(x) \to u(x)$ a.e. in Ω;
- there exists $w \in L^2(\Omega)$ such that $|u_k(x)| \le w(x)$ a.e. in Ω and for all k.

Notice now that since F is continuous we have $F(u_k(x)) \to F(u(x))$ a.e. in Ω, and due to the growth properties of F, we also have

$$|F(u_k(x))| \le M\,|u_k(x)| \le M\,w(x)$$

a.e. in Ω and for all k. Since Ω is bounded, $w \in L^1(\Omega)$, and by dominated convergence we obtain $F(u_k) \to F(u)$ in $L^1(\Omega)$; in particular,

$$\int_\Omega F(u_k)\,dx \to \int_\Omega F(u)\,dx.$$

We also have, of course,

$$\int_\Omega h u_k\,dx \to \int_\Omega h u\,dx \quad \text{and} \quad \|u\|^2 \le \liminf_k \|u_k\|,$$

by weak lower semicontinuity of the norm. Thus

$$
\begin{aligned}
I(u) &= \frac{1}{2}\|u\|^2 - \int_\Omega F(u)\,dx - \int_\Omega h u\,dx \\
&\le \liminf_k \frac{1}{2}\|u_k\|^2 - \lim_k \int_\Omega F(u_k)\,dx - \lim_k \int_\Omega h u_k\,dx \\
&= \liminf_k \left(\frac{1}{2}\|u_k\|^2 - \int_\Omega F(u_k)\,dx - \int_\Omega h u_k\,dx \right) = \liminf_k I(u_k) = m.
\end{aligned}
$$

But $u \in H_0^1(\Omega)$, so that $I(u) \ge m$, which shows that $I(u) = m$. Therefore u is a global minimum for I, and hence it is a critical point, namely a solution to (2.1). $\square$

Remark 2.1.4 Analyzing the preceding proof one sees that what we actually did is to show that I is coercive and weakly lower semicontinuous on $H_0^1(\Omega)$. These are exactly the assumptions that one needs in the (generalized) Weierstrass Theorem to deduce the existence of a global minimum, see Remark 1.5.7.

The boundedness of f in the previous result has been used to show that the nonlinear term $\int_\Omega F(u)\,dx$ does not destroy the growth properties of I inherited by the term $\|u\|^2$. This occurred because, as we have seen, the nonlinear term grows at most linearly. Now this is not really necessary: it is enough that this term grows *less than quadratically*. Let us see what kind of assumptions we can use in this sense in the next two results.

We begin by replacing the boundedness condition ($\mathbf{h_3}$) by the growth assumption

($\mathbf{h_4}$) $f : \mathbb{R} \to \mathbb{R}$ is continuous and there exist $\sigma \in (0, 1)$ and $a, b > 0$ such that

$$|f(t)| \leq a + b|t|^{\sigma} \quad \forall t \in \mathbb{R}.$$

Thus f is no longer bounded, but is allowed to grow *sublinearly* ($\sigma < 1$). It follows that F grows at most *subquadratically*, in the sense that for some $a_1, b_1 > 0$,

$$|F(t)| \leq a_1 + b_1|t|^{\sigma+1} \quad \forall t \in \mathbb{R}, \tag{2.4}$$

with $\sigma + 1 < 2$.

Theorem 2.1.5 *Under the assumptions* ($\mathbf{h_1}$), ($\mathbf{h_2}$) *and* ($\mathbf{h_4}$), *Problem* (2.1) *admits at least one solution.*

Proof Working as in the preceding proof we first show that I is coercive. Using the fact that $\sigma + 1 < 2$ we have

$$\left| \int_{\Omega} F(u) \, dx \right| \leq a_1 |\Omega| + b_1 \int_{\Omega} |u|^{\sigma+1} \, dx \leq C_1 + C_2 \|u\|^{1+\sigma},$$

thanks to the continuity of the embedding $H_0^1(\Omega) \hookrightarrow L^{\sigma+1}(\Omega)$. Then

$$I(u) = \frac{1}{2}\|u\|^2 - \int_{\Omega} F(u) \, dx - \int_{\Omega} hu \, dx \geq \frac{1}{2}\|u\|^2 - C_1 - C_2\|u\|^{\sigma+1} - |h|_2|u|_2$$

$$\geq \frac{1}{2}\|u\|^2 - C_2\|u\|^{\sigma+1} - C_3\|u\| - C_1,$$

and coercivity follows.

Let now $\{u_k\}_k \subset H_0^1(\Omega)$ be a minimizing sequence for I. As in the proof of Theorem 2.1.2 above, $\{u_k\}_k$ is bounded and therefore, up to subsequences, it converges weakly to some $u \in H_0^1(\Omega)$ and satisfies the same properties as in the preceding case. Then, reasoning as we did above, we obtain again

$$\int_{\Omega} F(u_k) \, dx \to \int_{\Omega} F(u) \, dx,$$

so that

$$I(u) \leq \liminf_k \left(\frac{1}{2}\|u_k\|^2 - \int_{\Omega} F(u_k) \, dx - \int_{\Omega} h \, u_k \, dx \right) = \liminf_k I(u_k) = \inf_{H_0^1(\Omega)} I.$$

The function u is a global minimum, hence a critical point of I, and we have found a solution of (2.1). $\qquad \square$

In our quest for more general assumptions we now try to go one step further: precisely, can we allow a *linear* growth for f, and then a *quadratic* growth for F? The answer is in the affirmative, provided we supply a quantitative control of the linear growth. This control is formulated in terms of the first eigenvalue $\lambda_1 = \lambda_1(-\Delta + q)$ in the following assumption.

($\mathbf{h_5}$) $f : \mathbb{R} \to \mathbb{R}$ is continuous and there exist $a > 0$ and $b \in (0, \lambda_1)$ such that

$$|f(t)| \le a + b|t| \quad \forall t \in \mathbb{R}.$$

Integrating, it follows immediately that

$$|F(t)| \le a|t| + \frac{b}{2}|t|^2 \quad \forall t \in \mathbb{R}.$$

Notice the difference with respect to (1.9): this is because we now want to keep the coefficient in front of $|t|^2$ as small as possible.

Theorem 2.1.6 *Under the assumptions* ($\mathbf{h_1}$), ($\mathbf{h_2}$) *and* ($\mathbf{h_5}$), *Problem* (2.1) *admits at least one solution.*

Proof To control the term $\int_\Omega F(u)\,dx$ we use the characterization of the first eigenvalue, Theorem 1.7.6. We have

$$\left| \int_\Omega F(u)\,dx \right| \le a \int_\Omega |u|\,dx + \frac{b}{2} \int_\Omega |u|^2\,dx \le C\|u\| + \frac{b}{2\lambda_1}\|u\|^2,$$

so that

$$I(u) = \frac{1}{2}\|u\|^2 - \int_\Omega F(u)\,dx - \int_\Omega hu\,dx \ge \frac{1}{2}\|u\|^2 - C\|u\| - \frac{1}{2}\frac{b}{\lambda_1}\|u\|^2 - |h|_2|u|_2$$

$$\ge \frac{1}{2}\left(1 - \frac{b}{\lambda_1}\right)\|u\|^2 - C_1\|u\|.$$

Since $b < \lambda_1$, the functional is coercive.

The remaining part of the proof works exactly as in the preceding theorems. □

Remark 2.1.7 In the literature, the growth conditions contained in assumptions ($\mathbf{h_4}$) and ($\mathbf{h_5}$) are often written

$$\limsup_{t \to \pm\infty} \frac{|f(t)|}{|t|^\sigma} < +\infty \quad \text{and} \quad \limsup_{t \to \pm\infty} \frac{|f(t)|}{|t|} < \lambda_1$$

respectively.

Remark 2.1.8 It is interesting to inspect what happens if we allow $b \ge \lambda_1$ in ($\mathbf{h_5}$). In this case the functional I is *no longer coercive* and may be unbounded from below. In some cases, as for example if we take $f(t) = \lambda_k t$ ($k \ge 1$), Problem (2.1) has *no solution* for some h (see Theorem 1.7.8). Later we will see how to deal with nonlinearities that grow more than quadratically.

We now examine a variant of Problem (2.1), with the aim of showing how the variational information can be of help in establishing existence results. Consider

$$\begin{cases} -\Delta u + q(x)u = f(u) & \text{in } \Omega, \\ u = 0 & \text{on } \partial\Omega. \end{cases} \tag{2.5}$$

If $f(0) = 0$, a frequent case in the applications, then the problem admits $u \equiv 0$ as a solution (called the trivial solution).

Without further assumptions, it may very well be that the trivial solution is the *only* solution. For example, if $f(t)t \leq 0$ for all t, then any weak solution satisfies

$$\|u\|^2 = \int_\Omega f(u)u \, dx \leq 0,$$

and hence $u \equiv 0$.

In the next result we show a condition that prevents this fact.

Theorem 2.1.9 *Let* ($\mathbf{h_1}$) *hold. Assume moreover that* $f : \mathbb{R} \to \mathbb{R}$ *is continuous and satisfies*

$$f(0) = 0 \quad and \quad \limsup_{t \to \pm\infty} \frac{|f(t)|}{|t|} < \lambda_1.$$

Then Problem (2.5) *admits at least one solution (which may be trivial).*

If in addition f *also satisfies*

$$\liminf_{t \to 0^+} \frac{f(t)}{t} > \lambda_1, \tag{2.6}$$

then Problem (2.5) *admits at least one nontrivial solution.*

Proof The first part is a special case of Theorem 2.1.6. We now show that under condition (2.6) the solution found in the first part is not identically zero. We use a *level argument*, as follows.

First notice that by (2.6), there exists $\beta > \lambda_1$ and $\delta > 0$ such that

$$f(t) \geq \beta t \quad \forall t \in [0, \delta],$$

which implies that

$$F(t) \geq \frac{1}{2} \beta t^2 \quad \forall t \in [0, \delta].$$

Let $\varphi_1 > 0$ be the first eigenfunction of $-\Delta + q(x)$, and take $\varepsilon > 0$ so small that $\varepsilon \varphi_1(x) < \delta$ for almost every x; this is possible because $\varphi_1 \in L^\infty(\Omega)$, see Theorem 1.7.3.

Then

$$F(\varepsilon\varphi_1(x)) \geq \frac{1}{2}\beta\varepsilon^2\varphi_1^2(x)$$

a.e. in Ω. This implies that

$$I(\varepsilon\varphi_1) = \frac{1}{2}\|\varepsilon\varphi_1\|^2 - \int_\Omega F(\varepsilon\varphi_1) \, dx \leq \frac{1}{2}\varepsilon^2\|\varphi_1\|^2 - \frac{1}{2}\beta\varepsilon^2 \int_\Omega \varphi_1^2 \, dx$$

$$= \frac{1}{2}\varepsilon^2\lambda_1 \int_\Omega \varphi_1^2 \, dx - \frac{1}{2}\beta\varepsilon^2 \int_\Omega \varphi_1^2 \, dx = \frac{\varepsilon^2}{2}(\lambda_1 - \beta) \int_\Omega \varphi_1^2 \, dx < 0,$$

since $\beta > \lambda_1$. Let u be the solution that minimizes I. Then

$$I(u) = \min_{v \in H_0^1(\Omega)} I(v) \leq I(\varepsilon\varphi_1) < 0.$$

As $I(0) = 0$, u cannot be the trivial solution. □

Remark 2.1.10 It is possible to show that the preceding problem admits a *nonnegative* solution. Indeed it is enough to proceed as in Example 1.7.10.

Since **(h₃)** implies **(h₄)** that implies **(h₅)**, it is clear Theorem 2.1.6 implies Theorem 2.1.5 that in turn implies Theorem 2.1.2. As a further example we examine now another case in which we can apply the scheme of the previous results and that leads to a theorem that is independent of the preceding ones. Consider the assumption

(h₆) $f : \mathbb{R} \to \mathbb{R}$ is continuous and there exist $a, b > 0$ such that

$$|f(t)| \leq a + b|t|^{2^*-1} \quad \forall t \in \mathbb{R}.$$

Moreover

$$f(t)t \leq 0 \quad \forall t \in \mathbb{R}.$$

By integration one easily sees that there exist $a_1, b_1 > 0$ such that

$$|F(t)| \leq a_1 + b_1|t|^{2^*} \quad \forall t \in \mathbb{R}$$

and that

$$F(t) \leq 0 \quad \forall t \in \mathbb{R}$$

Notice that $-F$ is allowed to have *critical* growth, but F is not. Moreover the sign condition $f(t)t \leq 0$ prevents, as we have seen, the existence of nontrivial solutions when $h \equiv 0$. In spite of this, Problem (2.1) is solvable.

Theorem 2.1.11 *Under the assumptions* **(h₁)**, **(h₂)** *and* **(h₆)**, *Problem (2.1) admits at least one solution.*

Proof By Example 1.3.20, the usual functional I is differentiable on $H_0^1(\Omega)$. Coercivity is simple consequence of the sign of F:

$$I(u) = \frac{1}{2}\|u\|^2 - \int_\Omega F(u)\,dx - \int_\Omega hu\,dx \geq \frac{1}{2}\|u\|^2 - |h|_2|u|_2 \geq \frac{1}{2}\|u\|^2 - C\|u\|.$$

The proof then proceeds exactly as in the previous theorems. □

Remark 2.1.12 This last existence result is similar to the one obtained in Theorem 1.6.6. However here we do not assume the monotonicity of f, so that the functional needs not be convex. This implies that we have to prove the weakly lower semicontinuity, as in the other theorems of this section, and we do not have a uniqueness result.

Remark 2.1.13 If $|F|$ grows more than critically, the functional I is no longer well defined on $H_0^1(\Omega)$, because a function in $H_0^1(\Omega)$ need not be in $L^p(\Omega)$ if $p > 2^*$. This means that the integral of $F(u)$ may be divergent for some u.

2.2 A min–max Theorem

In this section we deal with a much more complex result than those treated so far; the proof of the main theorem is quite long and can be omitted upon a first reading.

First, a heuristic motivation. The results in the previous section show, very roughly speaking, that if the nonlinearity f "does not interact" with the spectrum of the differential operator, then the procedure used for the linear problem (minimization) still works in the nonlinear case: the functional is coercive, bounded below, and has a global minimum.

Let us see what we mean by "does not interact", at least in a simplified setting. Suppose that f is differentiable on $\mathbb{R}$ and that

$$\sup_{t \in \mathbb{R}} |f'(t)| < \lambda_1.$$

This assumption implies ($\mathbf{h_5}$) and in particular implies that the closure of the range of f' is contained in $(-\lambda_1, \lambda_1)$. This is the property that makes the functional coercive (actually, as we have seen, it is enough that the property holds for large t; one can also check that $\sup_{t \in \mathbb{R}} f'(t) < \lambda_1$ works as well).

The situation changes dramatically, even in the linear case, if we allow that $f'(t)$ lies (for t large) between some eigenvalues of the differential operator. For example, if we take $f(t) = \lambda t$ with $\lambda > \lambda_1$, then the associated functional is no longer coercive, and it is unbounded below. Thus no minimization is possible.

In *some* cases however, certain ideas used in the previous section can be modified to obtain again an existence result. We now describe one of these cases, returning to an assumption that does not require f to be differentiable. We assume that f is defined on $\mathbb{R}$ and

($\mathbf{h_7}$) There exist an integer $\nu \geq 1$ and $\alpha, \beta \in \mathbb{R}$ such that

$$\lambda_\nu < \alpha \leq \frac{f(s) - f(t)}{s - t} \leq \beta < \lambda_{\nu+1} \quad \forall s, t \in \mathbb{R}.$$

Remark 2.2.1 Clearly ($\mathbf{h_7}$) implies that f is globally Lipschitz continuous. Of course if f is differentiable, then ($\mathbf{h_7}$) is equivalent to

$$\lambda_\nu < \alpha \leq f'(t) \leq \beta < \lambda_{\nu+1} \tag{2.7}$$

for all $t \in \mathbb{R}$: the closure of the range of f' lies between two eigenvalues.

Remark 2.2.2 We notice for further use that by direct integration we obtain the following growth properties for f and for its primitive F: there exist constants $c \in \mathbb{R}$ such that

- $\alpha t + c \leq f(t) \leq \beta t + c \; \forall t \geq 0,$
- $\beta t + c \leq f(t) \leq \alpha t + c \; \forall t \leq 0,$
- $\frac{1}{2}\alpha t^2 + ct \leq F(t) \leq \frac{1}{2}\beta t^2 + ct \; \forall t \in \mathbb{R}.$

We are going to prove the following result.

Theorem 2.2.3 *Under the assumptions* (**h₁**), (**h₂**) *and* (**h₇**), *Problem* (2.1), *namely*

$$\begin{cases} -\Delta u + q(x)u = f(u) + h(x) & \text{in } \Omega, \\ u = 0 & \text{on } \partial\Omega, \end{cases}$$

admits exactly one solution.

We will prove this result through a series of lemmas, where we always assume (**h₁**), (**h₂**) and (**h₇**). The point is to study the properties of the functional

$$I(u) = \frac{1}{2}\|u\|^2 - \int_\Omega F(u)\,dx - \int_\Omega hu\,dx,$$

which is defined and differentiable on $H_0^1(\Omega)$ in view of the growth conditions on f (Example 1.3.20).

The space $H_0^1(\Omega)$ is endowed with the scalar product (2.2) and the corresponding norm. We denote

$$X_1 = \text{span}\{\varphi_1, \ldots, \varphi_\nu\}, \quad \text{and} \quad X_2 = X_1^\perp = \text{cl}\{\text{span}\{\varphi_k \mid k \geq \nu + 1\}\}, \quad (2.8)$$

where φ_k is the eigenfunction associated to λ_k and "cl" denotes closure in $H_0^1(\Omega)$. By definition the subspaces X_1 and X_2 are orthogonal and $H_0^1(\Omega) = X_1 \oplus X_2$.

Remark 2.2.4 In the following, with a slight abuse of notation, we will write any $w \in H_0^1(\Omega)$ indifferently as $w = u + v$ or $w = (u, v)$, with $u \in X_1$ and $v \in X_2$. This simplifies the notation at various stages.

Lemma 2.2.5 *We have*

$$\int_\Omega u^2\,dx \geq \frac{1}{\lambda_\nu}\|u\|^2 \quad \forall u \in X_1, \quad (2.9)$$

and

$$\int_\Omega v^2\,dx \leq \frac{1}{\lambda_{\nu+1}}\|v\|^2 \quad \forall v \in X_2. \quad (2.10)$$

Proof With the notation of Remark 1.7.5, if $u \in X_1$ then

$$\int_\Omega u^2\,dx = \sum_{k=1}^\nu \alpha_k^2 = \sum_{k=1}^\nu \frac{\beta_k^2}{\lambda_k} \geq \frac{1}{\lambda_\nu}\sum_{k=1}^\nu \beta_k^2 = \frac{1}{\lambda_\nu}\|u\|^2,$$

while if $v \in X_2$, then

$$\int_\Omega v^2\,dx = \sum_{k=\nu+1}^\infty \alpha_k^2 = \sum_{k=\nu+1}^\infty \frac{\beta_k^2}{\lambda_k} \leq \frac{1}{\lambda_{\nu+1}}\sum_{k=1}^\nu \beta_k^2 = \frac{1}{\lambda_{\nu+1}}\|v\|^2. \qquad \square$$

This lemma contains the relevant information to deduce the coercivity properties of the functional I. Of course we do not expect coercivity on the subspace X_1, and indeed on X_1 we have the opposite behavior. The next lemma clarifies this.

We define a functional $J : X_1 \times X_2 \to \mathbb{R}$ by

$$J(u, v) = I(u + v).$$

Lemma 2.2.6 *For every $v \in X_2$, the functional $J(\cdot, v) : X_1 \to \mathbb{R}$ is anticoercive (recall that this means that $-J(\cdot, v)$ is coercive). For every $u \in X_1$, the functional $J(u, \cdot) : X_2 \to \mathbb{R}$ is coercive.*

Proof By the growth properties listed in Remark 2.2.2 we have, for every fixed $v \in X_2$,

$$
\begin{aligned}
J(u, v) = I(u + v) &= \frac{1}{2} \|u + v\|^2 - \int_\Omega F(u + v) \, dx - \int_\Omega h(u + v) \, dx \\
&\leq \frac{1}{2} \|u\|^2 + \frac{1}{2} \|v\|^2 - \frac{1}{2} \int_\Omega \alpha(u + v)^2 dx \\
&\quad - c \int_\Omega (u + v) \, dx - \int_\Omega h(u + v) \, dx \\
&\leq \frac{1}{2} \|u\|^2 - \frac{\alpha}{2} \int_\Omega u^2 \, dx + c_1 \|u\| + c_2,
\end{aligned}
$$

where the c_i's are constants that depend on v, f, h but not on u. Since $u \in X_1$, from Lemma 2.2.5 we obtain

$$
J(u, v) \leq \frac{1}{2} \|u\|^2 - \frac{1}{2} \frac{\alpha}{\lambda_\nu} \|u\|^2 + c_1 \|u\| + c_2 = \frac{1}{2} \left(1 - \frac{\alpha}{\lambda_\nu} \right) \|u\|^2 + c_1 \|u\| + c_2.
$$

As $\alpha > \lambda_\nu$, this proves the first part.

With the same argument, for every fixed $u \in X_1$ we have

$$
\begin{aligned}
J(u, v) &\geq \frac{1}{2} \|u\|^2 + \frac{1}{2} \|v\|^2 - \frac{1}{2} \int_\Omega \beta(u + v)^2 - c \int_\Omega (u + v) \, dx - \int_\Omega h(u + v) \, dx \\
&\geq \frac{1}{2} \|v\|^2 - \frac{\beta}{2} \int_\Omega v^2 \, dx + c_3 \|v\| + c_4,
\end{aligned}
$$

and applying Lemma 2.2.5 we conclude that

$$
\begin{aligned}
J(u, v) &\geq \frac{1}{2} \|v\|^2 - \frac{1}{2} \frac{\beta}{\lambda_{\nu+1}} \|v\|^2 + c_3 \|v\| + c_4 \\
&\geq \frac{1}{2} \left(1 - \frac{\beta}{\lambda_{\nu+1}} \right) \|v\|^2 + c_3 \|v\| + c_4,
\end{aligned}
$$

where the constants do not depend on v. Since $\beta < \lambda_{\nu+1}$, also the second part is proved. $\qquad \square$

The next property is crucial.

Lemma 2.2.7 *For every $u \in X_1$, the functional $J(u, \cdot) : X_2 \to \mathbb{R}$ is strictly convex and for every $v \in X_2$, the functional $J(\cdot, v) : X_1 \to \mathbb{R}$ is strictly concave.*

Proof We are going to check the convexity properties through Proposition 1.5.10. Fix $u \in X_1$ and consider the functional $v \mapsto J(u, v) = I(u + v)$. This functional is differentiable, and, denoting by $\partial_2 J(u, v)$ its differential at $v \in X_2$, we have for all $w \in X_2$,

$$\partial_2 J(u, v)w = I'(u + v)w = (v|w) - \int_\Omega f(u + v)w \, dx - \int_\Omega h \, w \, dx.$$

Therefore, for every $v_1, v_2 \in X_2$,

$$[\partial_2 J(u, v_1) - \partial_2 J(u, v_2)] (v_1 - v_2)$$
$$= (v_1|v_1 - v_2) - \int_\Omega f(u + v_1)(v_1 - v_2) \, dx - \int_\Omega h(v_1 - v_2) \, dx$$
$$\quad - (v_2|v_1 - v_2) + \int_\Omega f(u + v_2)(v_1 - v_2) \, dx + \int_\Omega h(v_1 - v_2) \, dx$$
$$= \|v_1 - v_2\|^2 - \int_\Omega (f(u + v_1) - f(u + v_2))(v_1 - v_2) \, dx.$$

Now the assumption

$$\frac{f(t) - f(s)}{t - s} \le \beta$$

implies that for all $t, s \in \mathbb{R}$ there results

$$(f(t) - f(s))(t - s) \le \beta(t - s)^2,$$

so that

$$\int_\Omega (f(u + v_1) - f(u + v_2))(v_1 - v_2) \, dx$$
$$= \int_\Omega (f(u + v_1) - f(u + v_2))((u + v_1) - (u + v_2)) \, dx$$
$$\le \beta \int_\Omega ((u + v_1) - (u + v_2))^2 \, dx = \beta \int_\Omega (v_1 - v_2)^2 \, dx \le \frac{\beta}{\lambda_{\nu+1}} \|v_1 - v_2\|^2,$$

by Lemma 2.2.5. Thus we obtain

$$[\partial_2 J(u, v_1) - \partial_2 J(u, v_2)] (v_1 - v_2) \ge \left(1 - \frac{\beta}{\lambda_{\nu+1}}\right) \|v_1 - v_2\|^2 > 0 \quad (2.11)$$

for all $v_1 \ne v_2$, because $\beta < \lambda_{\nu+1}$.

Proposition 1.5.10 applies to show that for every $u \in X_1$ the functional $J(u, \cdot)$ is strictly convex.

The argument to prove the concavity of $J(\cdot, v)$ is essentially the same. Fix $v \in X_2$ and denote by $\partial_1 J(u, v)$ the differential of the functional $u \mapsto J(u, v)$ at u. As above we obtain, for all $z \in X_1$,

$$\partial_1 J(u, v)z = (u|z) - \int_\Omega f(u + v)z \, dx - \int_\Omega hz \, dx,$$

so that

$$[\partial_1 J(u_1, v) - \partial_2 J(u_2, v)] (u_1 - u_2)$$

$$= (u_1 | u_1 - u_2) - \int_\Omega f(u_1 + v)(u_1 - u_2) \, dx - \int_\Omega h(u_1 - u_2) \, dx$$

$$- (u_2 | u_1 - u_2) + \int_\Omega f(u_2 + v)(u_1 - u_2) \, dx + \int_\Omega h(u_1 - u_2) \, dx$$

$$= \|u_1 - u_2\|^2 - \int_\Omega (f(u_1 + v) - f(u_2 + v)) (u_1 - u_2) \, dx.$$

Thus, by Lemma 2.2.5,

$$\int_\Omega (f(u_1 + v) - f(u_2 + v)) (u_1 - u_2) \, dx \geq \int_\Omega \alpha(u_1 - u_2)^2 \, dx \geq \frac{\alpha}{\lambda_v} \|u_1 - u_2\|^2,$$

and then

$$[\partial_1 J(u_1, v) - \partial_1 J(u_2, v)] (u_1 - u_2) \leq \left(1 - \frac{\alpha}{\lambda_v}\right) \|u_1 - u_2\|^2 < 0$$

for all $u_1 \neq u_2$, because $\alpha > \lambda_v$. This shows that the functional $u \mapsto J(u, v)$ is strictly concave. $\qquad\square$

The preceding lemmas allow us to minimize J over X_2.

Lemma 2.2.8 *For every $u \in X_1$ there exists a unique $w = w(u) \in X_2$ such that*

$$J(u, w) = \min_{v \in X_2} J(u, v).$$

Proof The function $v \mapsto J(u, v)$ is continuous, coercive and strictly convex. By Theorems 1.5.6 and 1.5.8, it has a unique global minimum. $\qquad\square$

This results makes it possible to define a functional $G : X_1 \to \mathbb{R}$ by

$$G(u) = \min_{v \in X_2} J(u, v) = J(u, w(u))$$

that we are now going to study.

Lemma 2.2.9 *The functional G is bounded above, namely*

$$\sup_{u \in X_1} G(u) < +\infty.$$

Proof If $\sup_{u \in X_1} G(u) = +\infty$, there exists a sequence $\{u_k\}_{k \in \mathbb{N}} \subset X_1$ such that $G(u_k) \to +\infty$ as $k \to +\infty$. By the growth properties of F, for every $u \in X_1$, we have

$$G(u) \leq J(u, 0) \leq |J(u, 0)| \leq \frac{1}{2} \|u\|^2 + \int_\Omega |F(u)| \, dx + \int_\Omega |h \, u| \, dx$$

$$\leq c_5 \|u\|^2 + c_6 \|u\|,$$

for some positive constants c_5, c_6. Since $G(u_k) \to +\infty$, this shows that $\|u_k\| \to \infty$. But J is anticoercive on X_1, and then

$$G(u_k) \le J(u_k, 0) \to -\infty,$$

contradicting the choice of the sequence $\{u_k\}$.   □

Lemma 2.2.10 *The number*

$$s = \sup_{u \in X_1} G(u)$$

is attained in X_1, namely there exists $u_* \in X_1$ such that $G(u_*) = s$.

Proof Let $\{u_k\}_{k \in \mathbb{N}} \subset X_1$ be a maximizing sequence for G on X_1, so that $G(u_k) \to s$. The argument in the preceding proof shows that $\{u_k\}$ is bounded in X_1. Since X_1 has finite dimension, there exists $u_* \in X_1$ such that, up to subsequences, $u_k \to u_*$ in X_1, and then also in $H_0^1(\Omega)$ (all norms are equivalent on X_1).

Thus, for every fixed $v \in X_2$, we obtain, as $k \to \infty$,

- $\|u_k + v\|^2 \to \|u_* + v\|^2$,
- $\int_\Omega F(u_k + v)\, dx \to \int_\Omega F(u_* + v)$,
- $\int_\Omega h(u_k + v)\, dx \to \int_\Omega h(u_* + v)\, dx$,

and therefore

$$J(u_k, v) \to J(u_*, v).$$

On the other hand, $J(u_k, v) \ge G(u_k) \to s$, so that

$$J(u_*, v) \ge s.$$

Since this holds for every $v \in X_2$, we see that

$$G(u_*) = \min_{v \in X_2} J(u_*, v) \ge s = \sup_{u \in X_1} G(u),$$

that is,

$$G(u_*) = s = \max_{u \in X_1} G(u).$$ □

Let $v_* = w(u_*)$ be the unique element in X_2 given by Lemma 2.2.8. Notice that $G(u_*) = J(u_*, v_*)$. The following property is determinant to conclude the proof of Theorem 2.2.3.

Lemma 2.2.11 *The element $(u_*, v_*) \in X_1 \times X_2$ is a "saddle point" for J, in the sense that*

$$J(u, v_*) \le J(u_*, v_*) \le J(u_*, v) \quad \forall u \in X_1, \ \forall v \in X_2.$$

Proof Since $J(u_*, v_*) = G(u_*) = \min_{v \in X_2} J(u_*, v)$, the inequality

$$J(u_*, v_*) \le J(u_*, v) \quad \forall v \in X_2$$

is trivial, and the proof amounts to establish the other inequality, namely

$$J(u, v_*) \leq J(u_*, v_*) \quad \forall u \in X_1. \tag{2.12}$$

Once more, convexity plays a fundamental role.

For every $t \in (0, 1)$ and for every $u \in X_1$, set

$$w_t = w((1-t)u_* + tu),$$

where as usual $w(\cdot)$ is given by Lemma 2.2.8. Since J is concave in $u \in X_1$, and u_* maximizes G, we have

$$
\begin{aligned}
G(u_*) \geq G((1-t)u_* + tu) &= J((1-t)u_* + tu, w_t) \\
&\geq (1-t)J(u_*, w_t) + tJ(u, w_t) \\
&\geq (1-t)G(u_*) + tJ(u, w_t),
\end{aligned}
$$

from which we see that for all $t \in (0, 1)$ and all $u \in X_1$,

$$G(u_*) \geq J(u, w_t). \tag{2.13}$$

For u fixed, the preceding inequality, together with the fact that J is coercive in $v \in X_2$, shows that the set $\{w_t \mid t \in (0, 1)\}$ is bounded in X_2. Therefore we can take a sequence $\{t_k\}_{k \in \mathbb{N}} \subset (0, 1)$ such that

$$t_k \to 0 \quad \text{and} \quad w_{t_k} \rightharpoonup \tilde{w} \quad \text{in } X_2.$$

Since the functional $v \mapsto J(u, v)$ is weakly lower semicontinuous (Theorem 1.5.3), we obtain

$$J(u, \tilde{w}) \leq \liminf_k J(u, w_{t_k}) \leq G(u_*) \tag{2.14}$$

by (2.13).

We now show that $\tilde{w} = v_*$; in this case the proof is complete since (2.14) reads

$$J(u, v_*) \leq G(u_*) = J(u_*, v_*), \tag{2.15}$$

and u is arbitrary in X_1.

By the concavity in $u \in X_1$ and the definition of w, for the same sequence as above we have

$$
\begin{aligned}
(1-t_k)J(u_*, w_{t_k}) &+ t_k J(u, w_{t_k}) \\
&\leq J((1-t_k)u_* + t_k u, w_{t_k}) \\
&= G((1-t_k)u_* + t_k u) = \inf_{v \in X_2} J((1-t_k)u_* + t_k u, v).
\end{aligned}
$$

This means that for every $v \in X_2$,

$$(1-t_k)J(u_*, w_{t_k}) + t_k J(u, w_{t_k}) \leq J((1-t_k)u_* + t_k u, v). \tag{2.16}$$

When $k \to +\infty$ we have $t_k \to 0$, so that

$$(1-t_k)u_* + t_k u \to u_*, \quad w_{t_k} \rightharpoonup \tilde{w},$$

and from (2.16) we obtain, by continuity in u and weak lower semicontinuity in v,

$$J(u_*, \tilde{w}) \leq J(u_*, v).$$

This holds for every $v \in X_2$, and therefore

$$J(u_*, \tilde{w}) = \inf_{v \in X_2} (u_*, v) = G(u_*).$$

By Lemma 2.2.8, there exists a unique $w \in X_2$ such that $J(u_*, w) = G(u_*)$, and this element is exactly the one that we called v_*. Summing up, we obtain

$$\tilde{w} = v_*,$$

so that (2.15) is true.   □

End of the proof of Theorem 2.2.3 We first show that $u_* + v_*$ is a critical point for I. For every $u \in X_1$ and every $t \in \mathbb{R}$, let

$$\gamma(t) = J(u_* + tu, v_*) = I(u_* + tu + v_*).$$

The function γ is differentiable and has a maximum point at $t = 0$, by the preceding lemma. Then

$$0 = \gamma'(0) = I'(u_* + v_*)u.$$

In the same way, if for $v \in X_2$ we define

$$\eta(t) = J(u_*, v_* + tv) = I(u_* + v_* + tv),$$

then we see that η is differentiable and has a minimum point at $t = 0$. So,

$$0 = \eta'(0) = I'(u_* + v_*)v.$$

Adding the two equations we obtain

$$I'(u_* + v_*)(u + v) = 0$$

for every $u \in X_1$ and every $v \in X_2$; since $H_0^1(\Omega) = X_1 \oplus X_2$, we conclude that $I'(u_* + v_*) = 0$. The existence of a solution to Problem (2.1) is proved.

We now turn to the uniqueness question.

Assume that u_1, u_2 are two solutions of (2.1), that is,

$$(u_i|\psi) - \int_\Omega f(u_i)\psi \, dx - \int_\Omega h\psi = 0$$

for $i = 1, 2$ and for all $\psi \in H_0^1(\Omega)$. We define a function $a \in L^\infty(\Omega)$ by

$$a(x) = \begin{cases} \frac{f(u_1(x)) - f(u_2(x))}{u_1(x) - u_2(x)} & \text{if } u_1(x) \neq u_2(x), \\ \alpha & \text{if } u_1(x) = u_2(x). \end{cases}$$

From ($\mathbf{h_7}$) we see that $\alpha \leq a(x) \leq \beta$ for a.e. $x \in \Omega$. Subtracting the equation for u_2 from the equation for u_1 and setting $w = u_1 - u_2$, we see that w satisfies

$$(w|\psi) - \int_\Omega (f(u_1) - f(u_2))\psi \, dx = 0$$

for all $\psi \in H_0^1(\Omega)$, that we can also write

$$(w|\psi) - \int_\Omega a(x)w\psi \, dx = 0. \tag{2.17}$$

We now write $w = w_1 + w_2$, with $w_i \in X_i$ and we recall from Lemma 2.2.5 that

$$\int_\Omega w_1^2 \, dx \geq \frac{1}{\lambda_\nu} \|w_1\|^2, \qquad \int_\Omega w_2^2 \, dx \leq \frac{1}{\lambda_{\nu+1}} \|w_2\|^2.$$

Choosing $\psi = w_1$ and then $\psi = w_2$ in (2.17) we obtain

$$\|w_1\|^2 = \int_\Omega a w_1^2 \, dx + \int_\Omega a w_1 w_2 \, dx \quad \text{and} \quad \|w_2\|^2 = \int_\Omega a w_2^2 \, dx + \int_\Omega a w_1 w_2 \, dx,$$

so that

$$\|w_1\|^2 = \int_\Omega a w_1^2 \, dx - \int_\Omega a w_2^2 \, dx + \|w_2\|^2 \geq \alpha \int_\Omega w_1^2 \, dx - \beta \int_\Omega w_2^2 \, dx + \|w_2\|^2$$

$$\geq \frac{\alpha}{\lambda_\nu} \|w_1\|^2 - \frac{\beta}{\lambda_{\nu+1}} \|w_2\|^2 + \|w_2\|^2,$$

that can be written

$$\left(1 - \frac{\alpha}{\lambda_\nu}\right) \|w_1\|^2 \geq \left(1 - \frac{\beta}{\lambda_{\nu+1}}\right) \|w_2\|^2.$$

Since $\alpha > \lambda_\nu$ and $\beta < \lambda_{\nu+1}$, this inequality implies that $\|w_1\| = \|w_2\| = 0$, namely $w = 0$ and thus $u_1 = u_2$. Uniqueness is verified, and the proof of the theorem is complete. $\qquad \square$

Remark 2.2.12 The proof of uniqueness uses the equation satisfied by critical points and various inequalities. A more "abstract" proof involves just convexity properties and works as follows. Assume for simplicity that I has a critical point at zero (translating if necessary) and a critical point at $u + v \in X_1 \oplus X_2$. Define $\varphi : \mathbb{R}^2 \to \mathbb{R}$ as

$$\varphi(s,t) = I(su + tv) = J(su, tv).$$

It follows immediately from the properties of J (Lemma 2.2.7) that φ is strictly concave in s and strictly convex in t. Just as easily, φ has a critical point at $(0, 0)$ and another one at $(1, 1)$. This is impossible. Indeed 0 is necessarily a strict global maximum for $s \mapsto \varphi(s, 0)$ and a strict global minimum for $t \mapsto \varphi(0, t)$, while 1 is a strict global maximum for $s \mapsto \varphi(s, 1)$, and a strict global minimum for $t \mapsto \varphi(1, t)$. Then

$$\varphi(0, 0) < \varphi(0, 1) < \varphi(1, 1) < \varphi(1, 0) < \varphi(0, 0),$$

a contradiction. Thus I cannot have more than one critical point.

Remark 2.2.13 One of the interesting aspects of the previous theorem is the *procedure* by which we have found a critical point. Indeed, we have first split the space orthogonally as $H_0^1(\Omega) = X_1 \oplus X_2$, according to convexity and concavity properties of the functional I; then, writing the generic element of $H_0^1(\Omega)$ as $u + v$, with $u \in X_1$ and $v \in X_2$, we have found a critical level s for I as

$$s = \max_{u \in X_1} \min_{v \in X_2} I(u + v).$$

This is a first example of a procedure that we will generalize in Chap. 4.

Remark 2.2.14 It is important to notice that many steps of the proof of the theorem work because of convexity (or concavity) properties of the functional I. These properties hold because of the rather strong assumption (**h7**), that rules the behavior of the nonlinearity f on the entire real line. For example, if f is differentiable and we require that it satisfies (2.7) for $|t|$ large only, then the convexity properties of I fail, and we cannot prove Theorem 2.2.3. This does not mean that we cannot find a solution to the problem, but certainly we cannot repeat the above proof. These cases will be dealt with in later chapters with stronger methods.

Remark 2.2.15 In assumption (**h7**) one cannot allow $\alpha = \lambda_\nu$ or $\beta = \lambda_{\nu+1}$. For example, as we have already pointed out, the *linear* equation $-\Delta u = \lambda_\nu u + h$ in $H_0^1(\Omega)$ does not admit a solution for *every* $h \in L^2(\Omega)$.

Remark 2.2.16 As a final remark we point out that if $h = 0$ in Theorem 2.2.3, then the problem admits only the trivial solution when $f(0) = 0$. Indeed in this case $u = 0$ is a critical point of I, and, by uniqueness, it is its *only* critical point.

2.3 Superlinear Problems and Constrained Minimization

Up to now we have only treated problems where the nonlinear term f has an at most linear growth. In case the nonlinearity grows faster than linearly, one speaks of *superlinear* problems. For these types of problems the techniques used so far do not work anymore; for example the functionals associated to these problems are generally unbounded from below and they present a lack of convexity or concavity properties that makes the arguments of the previous sections useless.

Actually for rather general f and h only partial results are known, though a quite rich theory has been developed. We begin to present here some of the simplest cases, in which h is identically zero and f is a power. Further results will be presented in later sections.

We take throughout this section a real number p such that

$$2 < p < 2^* = \frac{2N}{N-2},$$

and we search a function u that satisfies the *superlinear* and *subcritical* problem

$$\begin{cases} -\Delta u + q(x)u = |u|^{p-2}u & \text{in } \Omega, \\ u = 0 & \text{on } \partial\Omega. \end{cases} \tag{2.18}$$

Of course this problem admits the zero solution, which poses an extra difficulty. We will prove the existence of a nontrivial solution by two different methods, which can be extended to cover more general problems. The main result is the following.

Theorem 2.3.1 *Let $p \in (2, 2^*)$ and assume that* $(\mathbf{h_1})$ *holds. Then Problem* (2.18) *admits at least one nonnegative and nontrivial solution.*

From now on we tacitly assume that the hypotheses of Theorem 2.3.1 hold. For the role of $(\mathbf{h_1})$, see Remark 2.1.1.

The functional whose critical points are the (weak) solutions of (2.18) is $I : H_0^1(\Omega) \to \mathbb{R}$ defined by

$$I(u) = \frac{1}{2} \int_\Omega |\nabla u|^2 \, dx + \frac{1}{2} \int_\Omega q(x) u^2 \, dx - \frac{1}{p} \int_\Omega |u|^p \, dx = \frac{1}{2} \|u\|^2 - \frac{1}{p} |u|_p^p,$$

which is differentiable by the results of Example 1.3.20. We notice immediately that I is not bounded below, since for every $u \neq 0$ we have

$$I(tu) = \frac{t^2}{2} \|u\|^2 - \frac{t^p}{p} |u|_p^p \to -\infty$$

if $t \to +\infty$, because $p > 2$. Notice also that I is the difference of two strictly convex functionals.

2.3.1 Minimization on Spheres

In the first method that we present the key property is the *homogeneity* of the two terms in I; indeed the first term is positively homogeneous of degree 2, while the second is positively homogeneous of degree p. This difference of degrees of homogeneity will play a central role at various steps of the proof.

Since the functional I is unbounded below, no minimization is possible on the whole space $H_0^1(\Omega)$. The first step consists in getting rid of this unboundedness by *constraining* the functional on a suitable set where it becomes bounded below. The first choice (a second one will be presented in the next section) is a *sphere* of $L^p(\Omega)$.

If, for every $\beta > 0$, we set

$$\Sigma_\beta = \left\{ u \in H_0^1(\Omega) \;\middle|\; \int_\Omega |u|^p \, dx = \beta \right\},$$

we see that I restricted to Σ_β takes the form $I(u) = \frac{1}{2} \|u\|^2 - \frac{1}{p}\beta$, so that it is certainly bounded from below. Now minimizing I on Σ_β is equivalent to minimizing just the square of the norm, so we set

$$m_\beta = \inf_{u \in \Sigma_\beta} \|u\|^2.$$

We are going to show that m_β is attained by some function, and that this function gives rise to a solution of Problem (2.18).

Lemma 2.3.2 *For every $\beta > 0$, the level m_β is attained by a nonnegative function, namely there exists $u \in \Sigma_\beta$, $u(x) \geq 0$ a.e. in Ω, such that*

$$\|u\|^2 = m_\beta.$$

Proof Let $\{u_k\}_k \subset \Sigma_\beta$ be a minimizing sequence for $\|u\|^2$. Obviously the sequence $\{|u_k|\}_k$ is still a minimizing sequence in Σ_β (see the properties of H_0^1 in Sect. 1.2.1) and therefore we can assume from the beginning that $u_k(x) \geq 0$ a.e. in Ω and for all k. This minimizing sequence is of course bounded in $H_0^1(\Omega)$, so that, up to subsequences,

$$u_k \rightharpoonup u \quad \text{in } H_0^1(\Omega), \qquad u_k \to u \quad \text{in } L^p(\Omega), \quad \text{and}$$

$$u_k(x) \to u(x) \quad \text{a.e. in } \Omega.$$

We then obtain immediately, by weak lower semicontinuity of the norm,

$$\|u\|^2 \leq m_\beta,$$

together with

$$\int_\Omega |u|^p \, dx = \beta, \quad \text{and} \quad u(x) \geq 0 \quad \text{a.e. in } \Omega.$$

Thus $u \in \Sigma_\beta$, and this shows that $\|u\|^2 = m_\beta$. Notice also that $u \in \Sigma_\beta$ implies that u does not vanish identically. □

Remark 2.3.3 The preceding proof is very simple. The reader should retain from it that the key assumption is the *subcritical* growth condition $p < 2^*$. For such p's the embedding of $H_0^1(\Omega)$ into $L^p(\Omega)$ is *compact*, and this is what allows one to say that $u_k \to u$ in L^p, and eventually that $u \in \Sigma_\beta$. This is essential to conclude that u is a minimum in Σ_β.

In *critical* problems, namely when $p = 2^*$, the compactness of the embedding fails and the above argument breaks down. Even worse, some problems may have no nontrivial solution. This is the starting point of a line of research, begun with [14], that is still very active today. For some references see [26, 45].

Lemma 2.3.4 *Let u be a minimizer found with the previous lemma. Then u satisfies*

$$\int_\Omega \nabla u \cdot \nabla v \, dx + \int_\Omega q(x) u v \, dx = \frac{m_\beta}{\beta} \int_\Omega |u|^{p-2} u v \, dx \qquad (2.19)$$

for all $v \in H_0^1(\Omega)$.

Proof Although u minimizes the functional $N(u) = \|u\|^2$ on Σ_β, we cannot conclude that the differential of N vanishes at u, because Σ_β is not a vector space. Concretely, this means that we cannot compare the values of $N(u)$ and of $N(u+v)$, because $u + v$ will not belong to Σ_β, in general. We have to construct small "variations" of u that lie on Σ_β.

To this aim, fix $v \in H_0^1(\Omega)$. For $s \in \mathbb{R}$ small enough, say $s \in (-\varepsilon, \varepsilon)$, the function $u + sv$, is not identically zero. Therefore there exists $t : (-\varepsilon, \varepsilon) \to (0, +\infty)$ such that

$$\int_\Omega |t(s)(u + sv)|^p \, dx = \beta;$$

precisely,

$$t(s) = \left(\frac{\beta}{\int_\Omega |u + sv|^p \, dx} \right)^{1/p}.$$

Notice that the application $s \mapsto t(s)(u + sv)$ defines a *curve* on Σ_β that passes through u when $s = 0$. The function t is differentiable on $(-\varepsilon, \varepsilon)$, and

$$t'(s) = -\beta^{1/p} \left(\int_\Omega |u + sv|^p \, dx \right)^{-\frac{1}{p}-1} \int_\Omega |u + sv|^{p-2}(u + sv)v \, dx.$$

Then we have

$$t(0) = 1 \quad \text{and} \quad t'(0) = -\beta^{-1} \int_\Omega |u|^{p-2} uv \, dx.$$

We define $\gamma : (-\varepsilon, \varepsilon) \to \mathbb{R}$ as

$$\gamma(s) = N(t(s)(u + sv)) = \|t(s)(u + sv)\|^2.$$

Since $t(s)(u + sv) \in \Sigma_\beta$ for every $s \in (-\varepsilon, \varepsilon)$, the point $s = 0$ is a local minimum for γ. The function γ is differentiable and

$$\gamma'(s) = 2\big(t(s)(u + sv) \,\big|\, t'(s)(u + sv) + t(s)v\big),$$

so that

$$0 = \gamma'(0) = 2t(0)t'(0)\|u\|^2 + 2t^2(0)\,(u|v) = -2\frac{m_\beta}{\beta} \int_\Omega |u|^{p-2} uv \, dx + 2(u|v).$$

We have thus shown that

$$(u|v) = \frac{m_\beta}{\beta} \int_\Omega |u|^{p-2} uv \, dx,$$

for every $v \in H_0^1(\Omega)$, namely (2.19). $\square$

Remark 2.3.5 The reader with some knowledge of differential geometry will have noticed that the preceding proof amounts to a direct check of the fact that the differential of N at u vanishes on the tangent space to Σ_β at u, as is always the case for minimizers of differentiable functionals defined on differentiable manifolds. Indeed the previous result can be obtained "abstractly" through the Lagrange Theorem on constrained extrema: setting

$$G(u) = \int_\Omega |u|^p \, dx,$$

we see that our problem consists in minimizing N constrained on $G^{-1}(\beta) = \Sigma_\beta$. Now G is of class C^1 and for $u \in G^{-1}(\beta)$, we have that

$$G'(u)u = pG(u) = p\beta \neq 0.$$

Therefore, by the Implicit Function Theorem, $G^{-1}(\beta)$ is a C^1 manifold. The Lagrange Theorem then says that if u minimizes N on $G^{-1}(\beta)$, there exists $\lambda \in \mathbb{R}$ (the *Lagrange multiplier*) such that

$$N'(u) = \lambda G'(u).$$

This equation is precisely (2.19), with $\lambda = m_\beta / \beta$. We do not carry out a more general theory of constrained critical points; the interested reader can consult [2, 17].

End of the proof of Theorem 2.3.1 The last step consists in getting rid of the Lagrange multiplier m_β / β, and this is where homogeneity plays again a fundamental role.

Let u be the minimum of N over Σ_β. Set $u = cw$, with $c \in \mathbb{R}$ to be determined. By the previous lemma, w satisfies

$$c(w|v) = \frac{m_\beta}{\beta} c^{p-1} \int_\Omega |w|^{p-2} wv\, dx$$

for all $v \in H_0^1(\Omega)$. Choosing $c = (\beta/m_\beta)^{\frac{1}{p-2}}$, we see that w (is nonnegative and) satisfies

$$(w|v) = \int_\Omega |w|^{p-2} wv\, dx$$

for all $v \in H_0^1(\Omega)$, namely is a weak (nontrivial) solution of (2.18). $\square$

Remark 2.3.6 We have obtained a solution by minimizing N on Σ_β. It is tempting to constrain I or N on Σ_γ with $\gamma \neq \beta$ and repeat the argument. However this does not produce a new solution. Indeed it is very easy to see that

$$\frac{m_\beta}{\beta^{2/p}} = \frac{m_\gamma}{\gamma^{2/p}}$$

and that u is a minimum of I on Σ_β if and only if $v = (\gamma/\beta)^{1/p} u$ is a minimum of I on Σ_γ. These two functions give rise to the *same* solution of (2.18). For this reason one normally chooses $\beta = 1$.

2.3.2 Minimization on the Nehari Manifold

We present a second approach to the search of solutions to (2.18), still based on constrained minimization. This approach is slightly more complicated than minimization on spheres, but has the advantage that it does not require the nonlinearity to be homogeneous, and can thus be applied to a wider class of problems (under convenient assumptions). To illustrate it we concentrate again on Problem (2.18), and we recall that the functional associated to it is

$$I(u) = \frac{1}{2}\int_\Omega |\nabla u|^2\,dx + \frac{1}{2}\int_\Omega q(x)u^2\,dx - \frac{1}{p}\int_\Omega |u|^p\,dx = \frac{1}{2}\|u\|^2 - \frac{1}{p}|u|_p^p,$$

defined and differentiable on $H_0^1(\Omega)$. Recall that I is unbounded below; once again we restrict I to a suitable set in order to get rid of this problem. In the previous section we have constrained I on a sphere, now we use the set

$$\mathcal{N} = \{u \in H_0^1(\Omega) \mid u \neq 0, I'(u)u = 0\}$$
$$= \left\{u \in H_0^1(\Omega) \;\middle|\; u \neq 0, \|u\|^2 = \int_\Omega |u|^p\,dx\right\}. \tag{2.20}$$

This set is called the *Nehari manifold*, and indeed it can be proved, under certain assumptions, that it is a differential manifold diffeomorphic to the unit sphere of $H_0^1(\Omega)$, see [2] or [48]. We do not prove nor use these properties, but we confine ourselves to an "elementary" approach.

Remark 2.3.7 By definition, the Nehari manifold contains all the nontrivial critical points of I.

Notice that on $\mathcal{N}$ the functional I reads

$$I(u) = \left(\frac{1}{2} - \frac{1}{p}\right)\|u\|^2 = \left(\frac{1}{2} - \frac{1}{p}\right)\int_\Omega |u|^p\,dx.$$

This shows at once that I is coercive on $\mathcal{N}$, in the sense that if $\{u_k\}_k \subset \mathcal{N}$ satisfies $\|u_k\| \to \infty$, then $I(u_k) \to \infty$.

We define

$$m = \inf_{u\in\mathcal{N}} I(u),$$

and we show, through a series of lemmas, that m is attained by some $u \in \mathcal{N}$ which is a critical point of I *considered on the whole space* $H_0^1(\Omega)$, and therefore a solution to (2.18).

We begin with some basic properties of $\mathcal{N}$ and I.

Lemma 2.3.8 *The Nehari manifold is not empty.*

Proof For every not identically zero $u \in H_0^1$, one sees immediately that $tu \in \mathcal{N}$ for some $t > 0$. Indeed, $tu \in \mathcal{N}$ is equivalent to

$$\|tu\|^2 = \int_\Omega |tu|^p\,dx,$$

which is solved by

$$t = \left(\frac{\|u\|^2}{\int_\Omega |u|^p\,dx}\right)^{\frac{1}{p-2}} > 0. \qquad \square$$

Remark 2.3.9 Even under more general assumptions on the nonlinearity (now we are considering only the model case $s \to |s|^{p-2}s$) one can prove that for every $u \neq 0$, there exists a *unique* $t = t(u) > 0$ such that $t(u)u \in \mathcal{N}$. Thus one can define a map ψ from the unit sphere of $H_0^1(\Omega)$ to $\mathcal{N}$ as $\psi(u) = t(u)u$. In many concrete cases this map is the diffeomorphism between the unit sphere and the Nehari manifold that we mentioned above, see again [2] or [48].

Lemma 2.3.10 *We have*

$$m = \inf_{u \in \mathcal{N}} I(u) > 0.$$

Proof If $u \in \mathcal{N}$, by the Sobolev inequalities we have

$$\|u\|^2 = \int_{\Omega} |u|^p \, dx \leq C \|u\|^p,$$

for some $C > 0$. Since $\|u\| \neq 0$ and $p > 2$ we obtain

$$\|u\| \geq \left(\frac{1}{C}\right)^{\frac{1}{p-2}}, \tag{2.21}$$

for every $u \in \mathcal{N}$, so that

$$m = \inf_{u \in \mathcal{N}} I(u) = \left(\frac{1}{2} - \frac{1}{p}\right) \inf_{u \in \mathcal{N}} \|u\|^2 \geq \left(\frac{1}{2} - \frac{1}{p}\right)\left(\frac{1}{C}\right)^{\frac{1}{p-2}} > 0. \qquad \square$$

Lemma 2.3.11 *The level m is attained by a nonnegative function, namely there exists $u \in \mathcal{N}$, $u(x) \geq 0$ a.e. in Ω, such that*

$$I(u) = m.$$

Proof Let $\{u_k\}_k \subset \mathcal{N}$ be a minimizing sequence for I, namely such that $I(u_k) \to m$. Clearly $|u_k| \in \mathcal{N}$ and $I(|u_k|) = I(u_k)$, so that $\{|u_k|\}_k$ is another minimizing sequence; for this reason we assume straight away that $u_k(x) \geq 0$ a.e. in Ω for all k. We have already observed that I is coercive on $\mathcal{N}$; this implies that the sequence $\{u_k\}_k$ is bounded in $H_0^1(\Omega)$, and as usual this means that, up to subsequences,

$$u_k \rightharpoonup u \quad \text{in } H_0^1(\Omega), \qquad u_k \to u \quad \text{in } L^p(\Omega), \quad \text{and}$$

$$u_k(x) \to u(x) \quad \text{a.e. in } \Omega.$$

Then we have $u \geq 0$ a.e. and, by weak lower semicontinuity,

$$I(u) = \frac{1}{2}\|u\|^2 - \frac{1}{p}|u|_p^p \leq \liminf_k \left(\frac{1}{2}\|u_k\|^2 - \frac{1}{p}|u_k|_p^p\right)$$

$$= \liminf_k I(u_k) = m. \tag{2.22}$$

Since $u_k \in \mathcal{N}$, we have $\|u_k\|^2 = \int_\Omega |u_k|^p \, dx$. By (2.21) it cannot be $\|u_k\| \to 0$, and therefore $\int_\Omega |u_k|^p \, dx$ cannot tend to zero; thus, by strong convergence, $\int_\Omega |u|^p \, dx \neq 0$, which shows that $u \not\equiv 0$. Passing to the limit we obtain

$$\|u\|^2 \leq \int_\Omega |u|^p \, dx. \tag{2.23}$$

If $\|u\|^2 = \int_\Omega |u|^p \, dx$, then $u \in \mathcal{N}$ and (2.22) shows that u is the required minimizer. Since (2.23) holds, we only have to treat the case where

$$\|u\|^2 < \int_\Omega |u|^p \, dx. \tag{2.24}$$

We now show that if this happens, we reach a contradiction. Indeed, take $t > 0$ such that $tu \in \mathcal{N}$, namely

$$t = \left(\frac{\|u\|^2}{\int_\Omega |u|^p \, dx} \right)^{\frac{1}{p-2}}.$$

Since we are assuming (2.24), we deduce that $0 < t < 1$. But $tu \in \mathcal{N}$, so that

$$0 < m \leq I(tu) = \left(\frac{1}{2} - \frac{1}{p} \right) \|tu\|^2 = t^2 \left(\frac{1}{2} - \frac{1}{p} \right) \|u\|^2$$

$$\leq t^2 \liminf_k \left(\frac{1}{2} - \frac{1}{p} \right) \|u_k\|^2 = t^2 \liminf_k I(u_k) = t^2 m < m.$$

This is impossible, and the proof is complete. $\qquad\qquad\square$

Remark 2.3.12 Once again, the key property is the compactness of the embedding of $H_0^1(\Omega)$ into $L^p(\Omega)$, that has been used several times in the preceding proofs.

End of the proof of Theorem 2.3.1 Let $u \in \mathcal{N}$ be a minimizer for I found in the previous lemma. We show that $I'(u)v = 0$ for all $v \in H_0^1(\Omega)$, so that u is the required solution.

Notice that as in Lemma 2.3.4, we cannot conclude this directly because we have minimized I with the constraint $u \in \mathcal{N}$.

Take any $v \in H_0^1(\Omega)$. For every s in some small interval $(-\varepsilon, \varepsilon)$ certainly the function $u + sv$ does not vanish identically. Let $t(s) > 0$ be a function such that $t(s)(u + sv) \in \mathcal{N}$, namely

$$t(s) = \left(\frac{\|u + sv\|^2}{\int_\Omega |u + sv|^p \, dx} \right)^{\frac{1}{p-2}}.$$

The function $t(s)$ is a composition of differentiable functions, so it is differentiable; the precise expression of t' does not matter here. Notice also that $t(0) = 1$.

The map $s \mapsto t(s)(u + sv)$ defines a curve on $\mathcal{N}$ along which we evaluate I. Thus we define $\gamma : (-\varepsilon, \varepsilon) \to \mathbb{R}$ as

$$\gamma(s) = I(t(s)(u + sv)).$$

By construction, $s = 0$ is a minimum point for γ. Therefore

$$0 = \gamma'(0) = I'(t(0)u)[t'(0)u + t(0)v] = t'(0)I'(u)u + I'(u)v = I'(u)v.$$

For the last equality we have used the fact that $I'(u)u = 0$ because $u \in \mathcal{N}$. We have obtained $I'(u)v = 0$ for all $v \in H_0^1(\Omega)$, proving that u is a solution of (2.18). $\square$

Remark 2.3.13 The last part of the proof shows that a minimum of I *constrained* on the Nehari manifold $\mathcal{N}$ is actually a *free* critical point of I, on the whole space $H_0^1(\Omega)$. This remarkable fact is expressed by saying that the Nehari manifold is a *natural constraint* for I.

2.4 A Perturbed Problem

As we have anticipated the method of minimization on the Nehari manifold can be extended to cover more general problems than that of the preceding section.

We begin with an existence result for a *perturbed* problem, in the sense that we consider a power nonlinearity *plus* a fixed function $h \in L^2(\Omega)$. The result contained in this section is quite delicate, and can be omitted upon first reading.

We are going to prove that under suitable assumptions, and particularly if h is small, the problem admits *two* solutions. Let us make this more precise.

We consider, for $h \in L^2(\Omega)$ and $p \in (2, 2^*)$, the Dirichlet problem

$$\begin{cases} -\Delta u + q(x)u = |u|^{p-2}u + h(x) & \text{in } \Omega, \\ u = 0 & \text{on } \partial\Omega. \end{cases} \tag{2.25}$$

Or aim is to show that if $|h|_2$ is sufficiently small, then (2.25) admits at least two non-trivial solutions. This is not really surprising since the unperturbed problem ($h \equiv 0$) also admits two solutions: the one found with Theorem 2.3.1 and the trivial solution $u \equiv 0$. If h does not vanish identically, the trivial solution is replaced by a "true" nonzero solution. Thus one can think that for h small the two solutions are "perturbations" of the two solutions already present in the autonomous case.

We add that the solution corresponding in this scheme to the trivial solution of the unperturbed case can be found very easily, while most of the work must be devoted to the search of the analogue of the solution found in Theorem 2.3.1 by minimization on the Nehari manifold.

The functional associated to (2.25) is

$$J(u) = \frac{1}{2}\|u\|^2 - \frac{1}{p}\int_\Omega |u|^p\,dx - \int_\Omega hu\,dx = \frac{1}{2}\|u\|^2 - \frac{1}{p}|u|_p^p - \int_\Omega hu\,dx,$$

which is of course differentiable on $H_0^1(\Omega)$. The main result is the following.

Theorem 2.4.1 *Let $p \in (2, 2^*)$ and assume that* (**h₁**) *holds. Then there exists $\varepsilon^* > 0$ such that for every $h \in L^2(\Omega)$, with $|h|_2 \leq \varepsilon^*$, Problem (2.25) admits at least two solutions.*

We will prove this result going through a series of lemmas, as usual, and we start with the argument leading to the analogue of the solution found in Theorem 2.3.1, which is the hard part.

We introduce some notation. We denote by m the same constant as in the previous section, namely

$$m = \left(\frac{1}{2} - \frac{1}{p}\right) \inf_{u \in \mathcal{N}} \|u\|^2,$$

where $\mathcal{N}$ is the Nehari manifold defined in (2.20). Notice that $\mathcal{N}$ is *not* the Nehari manifold associated to J, but the "unperturbed" one, relative to the functional I of the previous section.

Next, we denote by S_p the best constant for the embedding of $H_0^1(\Omega)$ into $L^p(\Omega)$, that is,

$$S_p = \inf\{C > 0 \mid |u|_p \le C\|u\| \ \forall u \in H_0^1(\Omega)\}.$$

Lastly, we define the positive numbers

$$d_1 = \left(\frac{1}{(p-1)S_p^p}\right)^{\frac{1}{p-2}}, \qquad d_2 = \left(\frac{1}{2}d_1^2\right)^{1/p}, \qquad d_3 = \frac{1}{2}\min\left\{d_1, \frac{d_2}{S_p}\right\}.$$

Notice, for further reference, that $d_3 < d_1$.

Lemma 2.4.2 *There exists $\varepsilon_1 > 0$ such that for every $h \in L^2(\Omega)$ with $|h|_2 \le \varepsilon_1$ and for every $u \in H_0^1(\Omega)$, if*

$$\|u\|^2 = \int_\Omega |u|^p \, dx + \int_\Omega hu \, dx = |u|_p^p + \int_\Omega hu \, dx, \qquad (2.26)$$

then either

$$\|u\| < d_3 \quad or \quad \|u\| > d_1. \qquad (2.27)$$

If the second case occurs, we have also

$$|u|_p \ge d_2. \qquad (2.28)$$

Remark 2.4.3 Interpretation. Condition (2.26) expresses the fact that u belongs to the Nehari manifold relative to J (we will introduce it later). Then the meaning of the lemma is that, for $|h|_2$ small, if u is in the Nehari manifold, then either u is small ($\|u\| < d_3$) or u is large ($\|u\| > d_1$). The region $\|u\| \in [d_3, d_1]$ is forbidden.

Proof We begin by showing that if $|h|_2$ is small, one of the two inequalities in (2.27) must hold. Assuming (2.26), we obtain

$$\|u\|^2 \le S_p^p \|u\|^p + |h|_2 |u|_2 \le S_p^p \|u\|^p + c|h|_2 \|u\|,$$

where $c = \left(\frac{1}{\lambda_1}\right)^{1/2}$. Then, if $u \ne 0$,

$$\|u\| - S_p^p \|u\|^{p-1} - c|h|_2 \le 0. \qquad (2.29)$$

Consider now the function $\gamma : [0, +\infty) \to \mathbb{R}$ defined by

$$\gamma(t) = t - S_p^p t^{p-1} - c|h|_2.$$

Since

$$\gamma'(t) = 1 - (p-1)S_p^p t^{p-2},$$

the function γ has a unique global maximum point located at $t = d_1$. Moreover γ is strictly increasing on $(0, d_1)$, strictly decreasing on $(d_1, +\infty)$ and it satisfies $\gamma(0) < 0$ and $\lim_{t \to +\infty} \gamma(t) = -\infty$. With a simple computation one finds that

$$\gamma(d_1) = \frac{1}{S_p^{\frac{p}{p-2}}} \left(\frac{1}{p-1} \right)^{\frac{1}{p-2}} \frac{p-2}{p-1} - c|h|_2 =: \alpha_1 - c|h|_2.$$

Since $p > 2$, there results $\alpha_1 > 0$, and if we take $|h|_2 \leq \frac{\alpha_1}{2c}$, then

$$\gamma(d_1) \geq \frac{\alpha_1}{2} > 0.$$

This argument shows that if we assume

$$|h|_2 \leq \frac{\alpha_1}{2c},$$

then the function γ has exactly two zeros t_1, t_2 such that $t_1 < d_1 < t_2$, and $\gamma(t) > 0$ in (t_1, t_2), while $\gamma(t) < 0$ in $[0, t_1) \cup (t_2, +\infty)$.

We notice that t_1 satisfies

$$c|h|_2 = t_1 - S_p^p t_1^{p-1} = t_1 \left(1 - S_p^p t_1^{p-2} \right).$$

Since $t_1 < d_1$, we deduce from this and the definition of d_1 that

$$c|h|_2 \geq t_1 \left(1 - S_p^p d_1^{p-2} \right) = t_1 \left(1 - \frac{1}{p-1} \right) = t_1 \frac{p-2}{p-1},$$

so that

$$t_1 \leq \frac{p-1}{p-2} c|h|_2.$$

But then, if we take

$$|h|_2 < \frac{p-2}{p-1} \frac{d_3}{c},$$

we get $t_1 < d_3$.

Summing up, if we choose

$$|h|_2 < \min \left\{ \frac{p-2}{p-1} \frac{d_3}{c}, \frac{\alpha_1}{2c} \right\},$$

we obtain that $\gamma(t) \leq 0$ implies $t < d_3$ or $t > d_1$.

Therefore, with this choice of $|h|_2$, the inequality (2.29) implies (2.27) and thus also (2.26) implies (2.27). The first part is proved.

We still have to check (2.28). If (2.26) holds and $\|u\| > d_1$, we can write

$$|u|_p^p = \|u\|^2 - \int_\Omega hu\,dx \geq d_1^2 - |h|_2|u|_2 \geq d_1^2 - d|h|_2|u|_p,$$

where $d = |\Omega|^{\frac{p-2}{2p}}$. We obtain

$$|u|_p^p + d|h|_2|u|_p - d_1^2 \geq 0. \tag{2.30}$$

Consider the function $\eta : [0, +\infty) \to \mathbb{R}$ defined by

$$\eta(t) = t^p + d|h|_2 t - d_1^2.$$

We have

$$\eta(d_2) = \frac{1}{2}d_1^2 + d|h|_2 d_2 - d_1^2 = d|h|_2 d_2 - \frac{1}{2}d_1^2,$$

so that if

$$|h|_2 < \frac{d_1^2}{2dd_2},$$

there results $\eta(d_2) < 0$. Since $\eta'(t) > 0$ for all $t > 0$, we deduce that $\eta(t) < 0$ for $t \in [0, d_2]$, so that the inequality (2.30) implies $|u|_p \geq d_2$.

The proof is then complete, with the choice

$$\varepsilon_1 = \min\left\{\frac{p-2}{p-1}\frac{d_3}{c}, \frac{\alpha_1}{2c}, \frac{d_1^2}{2dd_2}\right\}. \qquad \square$$

From now on we always assume $|h|_2 < \varepsilon_1$; further restrictions will be imposed later.

We define the set

$$\mathcal{N}_h = \{u \in H_0^1(\Omega) \mid J'(u)u = 0, \|u\| > d_1\}$$
$$= \left\{u \in H_0^1(\Omega) \,\Big|\, \|u\|^2 = \int_\Omega |u|^p\,dx + \int_\Omega hu\,dx, \|u\| > d_1\right\}$$

and the value

$$m_h = \inf_{u \in \mathcal{N}_h} J(u).$$

Notice that $\mathcal{N}_h$ is not the complete Nehari manifold associated to J, but only a subset of it, containing "large" functions ($\|u\| > d_1$). On $\mathcal{N}_h$, the functional J takes the form

$$J(u) = \left(\frac{1}{2} - \frac{1}{p}\right)\|u\|^2 - \left(1 - \frac{1}{p}\right)\int_\Omega hu\,dx.$$

First we investigate under which conditions $\mathcal{N}_h$ is not empty.

Lemma 2.4.4 *There exists $\varepsilon_2 \in (0, \varepsilon_1]$ such that for every $h \in L^2(\Omega)$ with $|h|_2 \leq \varepsilon_2$, there results $\mathcal{N}_h \neq 0$.*

Proof Let $u \in H_0^1(\Omega) \setminus \{0\}$. We study the behavior of the function

$$t \mapsto J'(tu)tu = t^2\|u\|^2 - t^p \int_\Omega |u|^p \, dx - t \int_\Omega hu \, dx$$

$$= t\left[t\|u\|^2 - t^{p-1}|u|_p^p - \int_\Omega hu \, dx\right]$$

for $t > 0$ by analyzing the function

$$\gamma(t) = t\|u\|^2 - t^{p-1}|u|_p^p - \int_\Omega hu \, dx.$$

Since

$$\gamma'(t) = \|u\|^2 - (p-1)t^{p-2}|u|_p^p,$$

the function γ has a global maximum at

$$\tilde{t} = \left(\frac{\|u\|^2}{(p-1)|u|_p^p}\right)^{\frac{1}{p-2}}.$$

With an easy computation we see that

$$\gamma(\tilde{t}) = \frac{\|u\|^{2\frac{p-1}{p-2}}}{|u|_p^{\frac{p}{p-2}}}\alpha - \int_\Omega hu \, dx,$$

where

$$\alpha = \frac{1}{(p-1)^{\frac{1}{p-2}}}\frac{p-2}{p-1} > 0.$$

Then we obtain

$$\gamma(\tilde{t}) \geq \frac{\|u\|^{2\frac{p-1}{p-2}}}{\|u\|^{\frac{p}{p-2}}S_p^{\frac{p}{p-2}}}\alpha - \int_\Omega hu \, dx = \|u\|\frac{1}{S_p^{\frac{p}{p-2}}}\alpha - |h|_2|u|_2$$

$$\geq \|u\|\frac{1}{S_p^{\frac{p}{p-2}}}\alpha - c\|u\|\,|h|_2 = \|u\|\left(\frac{\alpha}{S_p^{\frac{p}{p-2}}} - c|h|_2\right).$$

If

$$|h|_2 \leq \frac{\alpha}{2cS_p^{\frac{p}{p-2}}},$$

we have $\gamma(\tilde{t}) > 0$. For such values of $|h|_2$, the function γ has then the following properties: $\gamma(t)$ is strictly increasing in $(0, \tilde{t})$, strictly decreasing in $(\tilde{t}, +\infty)$ and it satisfies $\gamma(\tilde{t}) > 0$ and $\lim_{t\to+\infty}\gamma(t) = -\infty$.

This shows that γ has at least one zero $t_1 \in (\tilde{t}, +\infty)$, which implies that the function $v = t_1 u$ satisfies (2.26). Moreover,

$$\|v\| = t_1\|u\| > \tilde{t}\|u\| = \left(\frac{\|u\|^2}{(p-1)|u|_p^p}\right)^{\frac{1}{p-2}}\|u\| \geq \frac{1}{(p-1)^{\frac{1}{p-2}}S_p^{\frac{p}{p-2}}} = d_1,$$

and this shows that $v \in \mathcal{N}_h$. The proof is then complete provided we choose

$$\varepsilon_2 = \min\left\{\varepsilon_1, \frac{\alpha}{2cS_p^{\frac{p}{p-2}}}\right\}.$$

$\square$

We now prove that for $|h|_2$ small, the levels m_h are uniformly bounded from above. A uniform bound from below will be obtained in Lemma 2.4.7.

Lemma 2.4.5 *Let $\varepsilon_3 = \min\{1, \varepsilon_2\}$. There exists $c_2 > 0$ such that for every $|h|_2 < \varepsilon_3$ there results $m_h \leq c_2$.*

Proof We denote by u_0 and m_0, respectively, the solution and the level of the solution of the unperturbed problem ($h \equiv 0$) solved in the preceding section, that is,

$$u_0 \in \mathcal{N}, \quad I(u_0) = \min_{u \in \mathcal{N}} I(u) = m_0.$$

From Lemma 2.4.4 we know that if $|h|_2 < \varepsilon_3$, there exists $t_h > 0$ such that $t_h u_0 \in \mathcal{N}_h$. Since $u_0 \in \mathcal{N}$, it satisfies $\|u_0\|^2 = |u_0|_p^p$, so that the condition $t_h u_0 \in \mathcal{N}_h$ is equivalent to

$$\left(t_h^2 - t_h^p\right)\|u_0\|^2 = t_h \int_\Omega h u_0 \, dx,$$

namely to

$$\left(t_h - t_h^{p-1}\right)\|u_0\|^2 = \int_\Omega h u_0 \, dx.$$

This last condition implies

$$\left(t_h - t_h^{p-1}\right)\|u_0\|^2 \geq -c|h|_2\|u_0\|,$$

i.e.

$$t_h - t_h^{p-1} \geq -\frac{c|h|_2}{\|u_0\|}.$$

If $|h|_2 < \varepsilon_3 \leq 1$ we obtain

$$t_h - t_h^{p-1} \geq -\frac{c}{\|u_0\|}. \tag{2.31}$$

Since the function $t \mapsto t - t^{p-1}$ tends to $-\infty$ as $t \to +\infty$, the inequality (2.31) implies the existence of $c_3 > 0$, independent of h, such that $t_h \leq c_3$. It is now simple to conclude. Since $t_h u_0 \in \mathcal{N}_h$ we have

$$
\begin{aligned}
m_h \leq J(t_h u_0) &= \left(\frac{1}{2} - \frac{1}{p}\right)\|t_h u_0\|^2 - \left(1 - \frac{1}{p}\right)\int_\Omega h t_h u_0 \, dx \\
&\leq \left(\frac{1}{2} - \frac{1}{p}\right)c_3^2\|u_0\|^2 + \left(1 - \frac{1}{p}\right)c_3\,c|h|_2\,\|u_0\| \\
&\leq \left(\frac{1}{2} - \frac{1}{p}\right)c_3^2\|u_0\|^2 + \left(1 - \frac{1}{p}\right)c_3\,c\,\|u_0\|.
\end{aligned}
$$

The last term of this inequality is a positive number that does not depend on h. We call it c_2 and the proof is complete. $\square$

We go on with a uniform bound on minimizing sequences.

Lemma 2.4.6 *There exists $c_4 > 0$ independent of h, such that if $|h|_2 < \varepsilon_3$, then there exists a minimizing sequence $\{u_k\}_k$ for m_h such that $\|u_k\| \leq c_4$ for all k, and hence also $|u_k|_p \leq S_p c_4$ for all k.*

Proof Let $|h|_2 < \varepsilon_3 \leq 1$ and let $\{v_k\}_k$ be a minimizing sequence for m_h, namely

$$v_k \in \mathcal{N}_h \quad \text{and} \quad J(v_k) \to m_h.$$

Since $m_h \leq c_2$, there exists $\tilde{k}$ such that for every $k \geq \tilde{k}$ there results $J(v_k) \leq 2c_2$, and this implies

$$
\begin{aligned}
2c_2 &\geq \left(\frac{1}{2} - \frac{1}{p}\right) \|v_k\|^2 - \left(1 - \frac{1}{p}\right) \int_\Omega h v_k \, dx \\
&\geq \left(\frac{1}{2} - \frac{1}{p}\right) \|v_k\|^2 - \left(1 - \frac{1}{p}\right) c |h|_2 \|v_k\| \\
&\geq a_1 \|v_k\|^2 - a_2 \|v_k\|,
\end{aligned}
$$

where $a_1 = (\frac{1}{2} - \frac{1}{p})$ and $a_2 = (1 - \frac{1}{p})c$. From this inequality one easily sees that

$$\|v_k\| \leq \frac{a_2 + \sqrt{a_2^2 + 8a_1 c_2}}{2a_1}.$$

Then it is enough to set $c_4 = \frac{a_2 + \sqrt{a_2^2 + 8a_1 c_2}}{2a_1}$ and $u_k = v_{\tilde{k}+k}$ to complete the proof. $\square$

The next result completes the estimate of Lemma 2.4.5.

Lemma 2.4.7 *There exists $\varepsilon_4 \leq \varepsilon_3$ such that if $|h|_2 < \varepsilon_4$, then $m_h \geq \frac{1}{2}m_0 > 0$.*

Proof In this proof we denote by J_h the functional J, to stress its dependence on h. We also recall that I is the functional associated to the "unperturbed" problem dealt with in the previous section and that $\mathcal{N}$ is the corresponding Nehari manifold.

For every h with $|h|_2 < \varepsilon_3$, we consider the minimizing sequence $\{u_k\}_k$ obtained in Lemma 2.4.6. Let t_k be such that $t_k u_k \in \mathcal{N}$ (notice that both u_k and t_k depend on h, but we do not denote it explicitly to have lighter notation).

As we know from the preceding section,

$$t_k = \left(\frac{\|u_k\|^2}{|u_k|_p^p}\right)^{\frac{1}{p-2}}.$$

On the other hand $u_k \in \mathcal{N}_h$, so that $\|u_k\|^2 = |u_k|_p^p + \int_\Omega h u_k \, dx$, and hence

$$t_k = \left(1 + \frac{\int_\Omega h u_k \, dx}{|u_k|_p^p}\right)^{\frac{1}{p-2}}. \tag{2.32}$$

Then we can write

$$m_0 \leq I(t_k u_k) = \left(\frac{1}{2} - \frac{1}{p}\right) t_k^2 \|u_k\|^2$$

$$= \left(\frac{1}{2} - \frac{1}{p}\right) t_k^2 \|u_k\|^2 - \left(1 - \frac{1}{p}\right) t_k^2 \int_\Omega h u_k \, dx + \left(1 - \frac{1}{p}\right) t_k^2 \int_\Omega h u_k \, dx$$

$$= t_k^2 J_h(u_k) + \left(1 - \frac{1}{p}\right) t_k^2 \int_\Omega h u_k \, dx. \tag{2.33}$$

We want to obtain a uniform bound from below on $J_h(u_k)$. First of all we estimate t_k from above. To this aim, by the estimates obtained in Lemmas 2.4.2 and 2.4.6 we have

$$\frac{|\int_\Omega h u_k \, dx|}{|u_k|_p^p} \leq |h|_2 \frac{c\|u_k\|}{|u_k|_p^p} \leq |h|_2 \frac{cc_4}{d_2^p},$$

so that if

$$|h|_2 \leq \frac{d_2^p}{cc_4} \left[\left(\frac{4}{3}\right)^{\frac{p-2}{2}} - 1\right],$$

we obtain

$$\frac{|\int_\Omega h u_k \, dx|}{|u_k|_p^p} \leq \left(\frac{4}{3}\right)^{\frac{p-2}{2}} - 1$$

and hence, from (2.32),

$$t_k \leq \left(\frac{4}{3}\right)^{1/2},$$

which is the desired estimate on t_k. Next we analyze the last term in (2.33). We see immediately that

$$\left|\left(1 - \frac{1}{p}\right) t_k^2 \int_\Omega h u_k \, dx\right| \leq \left(1 - \frac{1}{p}\right) t_k^2 |h|_2 c \|u_k\| \leq \frac{4}{3} |h|_2 \left(1 - \frac{1}{p}\right) cc_4.$$

We now choose

$$|h|_2 < \frac{m_0}{4(1 - \frac{1}{p})cc_4},$$

so that we obtain

$$\left|\left(1 - \frac{1}{p}\right) t_k^2 \int_\Omega h u_k \, dx\right| \leq \frac{m_0}{3}.$$

From this and from (2.33), we deduce that

$$t_k^2 J_h(u_k) \geq \frac{2}{3} m_0,$$

which implies first of all that $J_h(u_k) > 0$. Since we also have $t_k^2 \leq \frac{4}{3}$, we finally get to

$$\frac{2}{3} m_0 \leq t_k^2 J_h(u_k) \leq \frac{4}{3} J_h(u_k),$$

namely $\frac{1}{2}m_0 \le J_h(u_k)$. Passing to the limit as $k \to \infty$, we see that

$$\frac{1}{2}m_0 \le m_h.$$

All this holds with the bounds we described on $|h|_2$. The lemma is thus proved with

$$\varepsilon_4 = \min\left\{\varepsilon_3, \frac{d_2^p}{cc_4}\left[\left(\frac{4}{3}\right)^{\frac{p-2}{2}} - 1\right], \frac{m_0}{4(1 - \frac{1}{p})cc_4}\right\}. \qquad \square$$

We have now everything we need to conclude the argument. First of all we show that the minimum on $\mathcal{N}_h$ is attained.

Lemma 2.4.8 *There exists $\varepsilon_5 \le \varepsilon_4$ such that for every $|h|_2 < \varepsilon_5$ the infimum m_h is attained by some $u \in \mathcal{N}_h$.*

Proof To begin with, let $|h|_2 < \varepsilon_4$ and let $\{u_k\}$ be a minimizing sequence for m_h constructed as in Lemma 2.4.6. Being bounded, we can assume that, up to subsequences, there exists $u \in H_0^1$ such that

$$u_k \rightharpoonup u \quad \text{in } H_0^1(\Omega), \qquad u_k \to u \quad \text{in } L^p(\Omega) \text{ and in } L^2(\Omega).$$

It is then immediate to deduce that

$$J(u) \le \liminf_k J(u_k) = m_h, \tag{2.34}$$

and that

$$\|u\|^2 \le |u|_p^p + \int_\Omega hu\,dx. \tag{2.35}$$

Let us suppose first of all that in (2.35) equality holds, that is,

$$\|u\|^2 = |u|_p^p + \int_\Omega hu\,dx. \tag{2.36}$$

In Lemma 2.4.2 we have seen that if (2.36) holds, then either $\|u\| > d_1$ or $\|u\| < d_3$. In the first case, $u \in \mathcal{N}_h$, and then (2.34) implies that u is the minimum we are looking for. If, on the contrary, $\|u\| < d_3$, then recalling the definition of d_3 we obtain

$$|u|_p \le S_p\|u\| \le S_p d_3 < S_p \frac{d_2}{S_p} = d_2.$$

But Lemma 2.4.2 says that $|u_k|_p \ge d_2$, and, by the strong convergence in $L^p(\Omega)$ we obtain $|u|_p \ge d_2$. This contradiction shows that if (2.36) holds, it cannot be $\|u\| < d_3$ and hence, as we have seen, u is a minimum.

Let us suppose now that in (2.35) the strict inequality holds, namely

$$\|u\|^2 < |u|_p^p + \int_\Omega hu\,dx. \tag{2.37}$$

We know from Lemma 2.4.4 that there exists $t^* > 0$ such that $t^* u \in \mathcal{N}_h$ and

$$t^* > \tilde{t} = \left(\frac{\|u\|^2}{(p-1)|u|_p^p} \right)^{\frac{1}{p-2}}.$$

Since (2.35) holds, we have

$$\tilde{t} \leq \left(\frac{|u|_p^p + \int_\Omega hu\,dx}{(p-1)|u|_p^p} \right)^{\frac{1}{p-2}} = \left(\frac{1}{p-1} + \frac{\int_\Omega hu\,dx}{(p-1)|u|_p^p} \right)^{\frac{1}{p-2}}.$$

But on the other hand,

$$\frac{|\int_\Omega hu\,dx|}{(p-1)|u|_p^p} \leq |h|_2 c \frac{\|u\|}{(p-1)|u|_p^p} \leq |h|_2 cc_4 \frac{1}{(p-1)d_2^p}.$$

Since $\frac{1}{p-1} < 1$, it is clear that we can choose $\varepsilon_5 = \min\{ \frac{(p-2)(p-1)d_2^p}{2cc_4}, \varepsilon_4 \}$, to obtain that if $|h|_2 < \varepsilon_5$, then $\tilde{t} < 1$.

Let us consider now the function

$$\gamma(t) = t\|u\|^2 - t^{p-1}|u|_p^p - \int_\Omega hu\,dx$$

that we have already studied. We know that γ is decreasing in $(\tilde{t}, +\infty)$. Since $t^* u \in \mathcal{N}_h$, we have $\gamma(t^*) = 0$. The inequality (2.37) is equivalent to $\gamma(1) < 0$. Since $\tilde{t} < 1$ and $\tilde{t} < t^*$, we must necessarily have $t^* < 1$. This last estimate allows us to conclude. Indeed, since $t^* u \in \mathcal{N}_h$ we obtain

$$m_h \leq J(t^* u) = (t^*)^2 \left(\frac{1}{2} - \frac{1}{p} \right) \|u\|^2 - t^* \left(1 - \frac{1}{p} \right) \int_\Omega hu\,dx$$

$$\leq (t^*)^2 \liminf_k \left(\frac{1}{2} - \frac{1}{p} \right) \|u_k\|^2 - t^* \lim_k \left(1 - \frac{1}{p} \right) \int_\Omega hu_k\,dx$$

$$\leq t^* \liminf_k \left[\left(\frac{1}{2} - \frac{1}{p} \right) \|u_k\|^2 - \left(1 - \frac{1}{p} \right) \int_\Omega hu_k\,dx \right]$$

$$= t^* \liminf_k J(u_k) = t^* m_h < m_h.$$

The last strict inequality holds because $t^* \in (0, 1)$ and $m_h > 0$. The contradiction shows that (2.37) cannot hold, so that (2.36) is true. In this case, as we have seen, u is the required minimum. $\qquad\square$

The last step consists in proving that the minimum is a critical point of J on the whole space $H_0^1(\Omega)$.

Lemma 2.4.9 *There exists $\varepsilon_6 \in (0, \varepsilon_5)$ such that if $|h|_2 < \varepsilon_6$, then the minimum u of J on $\mathcal{N}_h$ satisfies $J'(u)v = 0$ for all $v \in H_0^1(\Omega)$.*

Proof Fix $v \in H_0^1(\Omega)$ and consider the function $\varphi : \mathbb{R} \times (0, +\infty) \to \mathbb{R}$ defined by

$$\varphi(s, t) = t^2 \|u + sv\|^2 - t^p |u + sv|^p - t \int_\Omega h(u + sv)\,dx.$$

Since $u \in \mathcal{N}_h$ we have $\varphi(0, 1) = 0$. Clearly φ is of class C^1 and

$$\frac{\partial \varphi}{\partial t}(0, 1) = 2\|u\|^2 - p|u|_p^p - \int_\Omega hu\,dx = (2 - p)\|u\|^2 + (p - 1)\int_\Omega hu\,dx.$$

If it were $\frac{\partial \varphi}{\partial t}(0, 1) = 0$, then we would have

$$\|u\|^2 = \frac{p-1}{p-2}\int_\Omega hu\,dx \le \frac{p-1}{p-2}|h|_2 c\|u\|,$$

namely,

$$\|u\| \le \frac{p-1}{p-2}|h|_2 c. \tag{2.38}$$

However, we know that $\|u\| > d_1$. If we take

$$|h|_2 < \frac{p-2}{c(p-1)}d_1,$$

then (2.38) yields $\|u\| < d_1$, a contradiction. Therefore, for such choices of h it must be $\frac{\partial \varphi}{\partial t}(0, 1) \ne 0$. We can then apply the Implicit Function Theorem, obtaining that there exist a number $\delta > 0$ and a C^1 function $t(s) : (-\delta, \delta) \to \mathbb{R}$ such that $\varphi(s, t(s)) = 0$ for every $s \in (-\delta, \delta)$, and $t(0) = 1$. Since $\|u\| > d_1$, we can also take δ so small that $t(s)(u + sv) > d_1$. We have thus constructed a differentiable curve $s \mapsto \mathcal{N}_h$ passing through u when $s = 0$. We now evaluate J along this curve, by considering the function

$$\gamma(s) = J(t(s)(u + sv)).$$

The function γ is differentiable and has a local minimum at $s = 0$. Then

$$0 = \gamma'(0) = J'(u)[t'(0)u + t(0)v] = t'(0)J'(u)u + J'(u)v = J'(u)v,$$

because $u \in \mathcal{N}_h$. Since this holds for every $v \in H_0^1(\Omega)$, it is enough to take a positive ε_6 such that

$$\varepsilon_6 < \min\left\{\varepsilon_5, \frac{(p-2)d_1}{c(p-1)}\right\}$$

to conclude. $\square$

We have found a solution to Problem (2.25). We now complete the proof of the theorem by constructing a second solution.

Lemma 2.4.10 *For every $\varepsilon > 0$ there exists $\delta > 0$ such that if $|h|_2 < \delta$, then Problem* (2.25) *admits a solution u_h satisfying $\|u_h\| < \varepsilon$.*

Proof Let as usual $I(u) = \frac{1}{2}\|u\|^2 - \frac{1}{p}|u|_p^p$; we have

$$I(u) \ge \frac{1}{2}\|u\|^2 - \frac{S_p^p}{p}\|u\|^p.$$

The function

$$g(t) = \frac{1}{2}t^2 - \frac{S_p^p}{p}t^p$$

is strictly increasing in a right neighborhood of 0, is continuous and satisfies $g(0) = 0$; therefore there certainly exists $\varepsilon' \le \varepsilon$ such that for all $t \in (0, \varepsilon')$ there results $g(t) > 0$.

Then fixing any $\eta \in (0, \varepsilon')$, we obtain $I(u) \ge g(\eta) > 0$ if $\|u\| = \eta$. We deduce that

$$J(u) = I(u) - \int_\Omega hu\,dx \ge g(\eta) - |h|_2 c\eta.$$

Choosing

$$\delta = \frac{g(\eta)}{2c\eta} \quad \text{and} \quad |h|_2 < \delta,$$

we conclude that, for $\|u\| = \eta$,

$$J(u) \ge \frac{g(\eta)}{2} > 0.$$

Let now

$$B_\eta = \{u \in H_0^1(\Omega) \mid \|u\| \le \eta\} \quad \text{and} \quad n_\eta = \inf_{u \in B_\eta} J(u).$$

Clearly $n_\eta \ne \pm\infty$. Moreover $n_\eta \le J(0) = 0$. It is then straightforward to check, with arguments already used many times in the first part of this chapter, that the value n_η is attained by some $u_h \in B_\eta$. Since $J(u_h) = n_\eta \le 0$, it cannot be $\|u_h\| = \eta$, and therefore u_h lies in the interior of the ball B_η. The function u_h is thus a local minimum for J and solves (2.25). Moreover, $\|u_h\| < \varepsilon$.   □

End of the proof of Theorem 2.4.1 Thanks to Lemma 2.4.9 we know that for every $|h|_2 < \varepsilon_6$, there exists a solution u_1 of Problem (2.25) with $\|u_1\| > d_1$. By Lemma 2.4.10, choosing $\varepsilon = d_1$, we can fix $\delta > 0$ such that for every $|h|_2 < \delta$ there exists a solution u_h of Problem (2.25) with $\|u_h\| < d_1$. If

$$|h|_2 < \varepsilon^* := \min\{\varepsilon_6, \delta\},$$

we obtain two *distinct* solutions of Problem (2.25). The proof is complete.   □

2.5 Nonhomogeneous Nonlinearities

The method of minimization on the Nehari Manifold can also be extended to cases where the nonlinearity is not homogeneous, under suitable assumptions. In this section we provide some examples in this sense.

We begin with the following problem: find u such that

$$\begin{cases} -\Delta u + q(x)u = f(u) & \text{in } \Omega, \\ u = 0 & \text{on } \partial\Omega, \end{cases} \tag{2.39}$$

where the nonlinearity f satisfies the following assumptions. We recall that we denote $F(t) = \int_0^t f(s)ds$.

(f_1) $f : \mathbb{R} \to \mathbb{R}$ is of class C^1 and is odd.
(f_2) $f'(0) = 0$ and there exists $p \in (2, 2^*)$ such that $\limsup_{t \to +\infty} \frac{|f'(t)|}{t^{p-2}} < +\infty$.
(f_3) There exists $\mu > 2$ such that $f(t)t \geq \mu F(t)$ for all $t \in \mathbb{R}$.
(f_4) For every $t \in (0, +\infty)$ there results $f'(t)t > f(t)$.

Remark 2.5.1 Notice that the power nonlinearity $f(u) = |u|^{p-2}u$, with $p \in (2, 2^*)$, treated in Sect. 2.3 satisfies all the above assumptions.

Actually, assumptions (f_3) and (f_4) imply that the functions

$$t \mapsto \frac{F(t)}{t^\mu} \quad \text{and} \quad t \mapsto \frac{f(t)}{t}$$

are increasing, as one checks immediately by differentiation. This provides some bounds at zero and at infinity. For example, comparing with $t = 1$ one finds that $F(t) \leq F(1)t^\mu$ for $t \in [0, 1]$ and $F(t) \geq F(1)t^\mu$ for $t \geq 1$, and similarly for f.

We will prove the following result.

Theorem 2.5.2 *Assume that* (h_1) *and* (f_1)–(f_4) *hold. Then Problem* (2.39) *admits at least one nontrivial and nonnegative solution.*

As usual, the proof will be split in a series of lemmas, in each of which the assumptions of Theorem 2.5.2 will be taken for granted.

We begin with the description of some properties of $f(t)$ and $F(t)$ that can be deduced from the assumptions (f_1)–(f_4) and that will be used during the proof.

Lemma 2.5.3 *We have*

(1) $\lim_{t \to 0} \frac{F(t)}{t^2} = 0$.
(2) *There exist positive constants M_1, M_2 such that*

$$|f'(t)| \leq M_1 |t|^{p-2} \quad \text{for } |t| > M_2.$$

(3) *There exist positive constants M_3, M_4 such that*

$$|F(t)| + |f(t)t| \leq M_3|t|^p \quad \text{for } |t| > M_4.$$

(4) *For every $\varepsilon > 0$ there exists $C_\varepsilon > 0$ such that*

$$|F(t)| + |f(t)t| \leq \varepsilon t^2 + C_\varepsilon|t|^p \quad \forall t \in \mathbb{R}. \tag{2.40}$$

(5) *There exists a positive constant D such that for every $t \geq 1$,*

$$f(t)t \geq Dt^\mu \quad \text{and} \quad F(t) \geq Dt^\mu. \tag{2.41}$$

Proof The first statement follows by applying twice the de l'Hôpital rule. Point (2) is a direct consequence of (f_2) and the oddness of f. Integrating, one obtains (3).

Likewise, the fact that $f'(0) = 0$ and $(\mathbf{f_2})$ imply (4), after integration. Lastly, (5) follows from Remark 2.5.1 and $(\mathbf{f_3})$.   □

Weak solutions to Problem (2.39) are critical points of the functional

$$I(u) = \frac{1}{2}\|u\|^2 - \int_\Omega F(u)\,dx$$

where the norm is the same as in Sect. 2.3. The Nehari manifold in the present case is

$$\mathcal{N} = \{u \in H_0^1(\Omega) \mid I'(u)u = 0, u \neq 0\}$$
$$= \left\{u \in H_0^1(\Omega) \,\Big|\, \|u\|^2 = \int_\Omega f(u)u\,dx, u \neq 0\right\}.$$

As usual, we begin by checking that working on the Nehari manifold makes sense.

Lemma 2.5.4 *The Nehari manifold $\mathcal{N}$ is not empty.*

Proof We prove that for every $u \in H_0^1(\Omega)$, $u \neq 0$, there exists $t > 0$ such that $tu \in \mathcal{N}$. We begin by assuming that $u(x) \geq 0$ a.e., and we consider the function

$$\gamma(t) = I'(tu)tu = t^2\|u\|^2 - \int_\Omega f(tu)tu\,dx.$$

We want to study the behavior of γ for $t \to 0^+$ and for $t \to +\infty$. The aim is to find a zero of γ.

Let

$$\varphi(t) = \int_\Omega f(tu(x))tu(x)\,dx.$$

By (2.40), for every $\varepsilon > 0$ there exists $C_\varepsilon > 0$ such that

$$|f(tu(x))tu(x)| \leq \varepsilon t^2 u(x)^2 + C_\varepsilon t^p u(x)^p \quad \text{for all } t > 0 \text{ and for a.e. } x \in \Omega.$$

Therefore

$$\varphi(t) = \left|\int_\Omega f(tu(x))tu(x)\,dx\right| \leq \varepsilon t^2 \int_\Omega u(x)^2\,dx + C_\varepsilon t^p \int_\Omega u(x)^p\,dx.$$

Since u is fixed, ε is arbitrary and $p > 2$, this shows that $\varphi(t) = o(t^2)$ as $t \to 0^+$. We have thus proved that

$$\gamma(t) = t^2\|u\|^2 + o(t^2) \quad \text{as } t \to 0^+.$$

Next we study γ for t large. We begin by choosing t_0 so large that the set

$$A = \{x \in \Omega \mid t_0 u(x) \geq 1\}$$

has positive measure; this is possible since $u(x) \geq 0$ a.e. and $u \neq 0$. Notice that if $t \geq t_0$, then $tu(x) \geq 1$ a.e. on A. Now we apply (2.41) with $t \geq t_0$, also recalling that f is positive on positive arguments:

$$\int_\Omega f(tu(x))tu(x)\,dx = \int_A f(tu(x))tu(x)\,dx + \int_{\Omega\setminus A} f(tu(x))tu(x)\,dx$$

$$\geq Dt^\mu \int_A u(x)^\mu\,dx.$$

Notice that (3) and (5) of Lemma 2.5.3 imply that $\mu \leq p$, so that the last integral is certainly finite.

The preceding inequality shows that

$$\gamma(t) \leq t^2\|u\|^2 - Dt^\mu \int_A u(x)^\mu\,dx \quad \text{for every } t \geq t_0.$$

The two properties of γ that we have established tell us that $\gamma(t)$ is strictly positive for positive and small values of t, while $\gamma(t) \to -\infty$ if $t \to +\infty$.

Then there exists $\tilde{t} > 0$ such that $\gamma(\tilde{t}) = 0$. Since $\tilde{t}u \not\equiv 0$, we conclude that $\tilde{t}u \in \mathcal{N}$.

Finally, if u is not almost everywhere positive, we argue on the function $v = |u|$: we find again $\tilde{t} > 0$ such that

$$\|\tilde{t}v\|^2 - \int_\Omega f(\tilde{t}v)\tilde{t}v\,dx = 0.$$

Since $f(t)t$ is even, of course

$$\int_\Omega f(\tilde{t}v)\tilde{t}v\,dx = \int_\Omega f(\tilde{t}u)\tilde{t}u\,dx \quad \text{and} \quad \|\tilde{t}v\|^2 = \|\tilde{t}u\|^2,$$

so that also $\|\tilde{t}u\|^2 - \int_\Omega f(\tilde{t}u)\tilde{t}u\,dx = 0$, and $\tilde{t}u \in \mathcal{N}$. $\square$

We can now define a candidate critical level as

$$m = \inf_{u\in\mathcal{N}} I(u)$$

and we try to see if m is attained on $\mathcal{N}$.

Lemma 2.5.5 *There results*

$$\inf_{u\in\mathcal{N}} \|u\|^2 > 0.$$

Proof Let λ_1 be the first eigenvalue of the operator $-\Delta + q(x)$, and choose a number $C_1 > 0$ such that

$$|f(t)t| \leq \frac{1}{2}\lambda_1 t^2 + C_1|t|^p \quad \forall t;$$

this is possible by (2.40).

For every $u \in \mathcal{N}$ we have

$$\|u\|^2 = \int_\Omega f(u)u\,dx \leq \frac{\lambda_1}{2}\int_\Omega u^2\,dx + C_1\int_\Omega |u|^p\,dx \leq \frac{1}{2}\|u\|^2 + C_2\|u\|^p,$$

so that

$$\|u\| \geq \left(\frac{1}{2C_2}\right)^{\frac{1}{p-2}},$$

which proves the claim.   □

Lemma 2.5.6 *The functional* I *is coercive on* $\mathcal{N}$ *and* $m > 0$.

Proof For every $u \in \mathcal{N}$, by (**f₃**), we have

$$I(u) = \frac{1}{2}\|u\|^2 - \int_\Omega F(u)\,dx = \left(\frac{1}{2} - \frac{1}{\mu}\right)\|u\|^2 + \frac{1}{\mu}\|u\|^2 - \int_\Omega F(u)\,dx$$

$$= \left(\frac{1}{2} - \frac{1}{\mu}\right)\|u\|^2 + \int_\Omega \left(\frac{1}{\mu}f(u)u - F(u)\right)dx \geq \left(\frac{1}{2} - \frac{1}{\mu}\right)\|u\|^2,$$

which shows that I is coercive on $\mathcal{N}$ and that

$$m = \inf_{u\in\mathcal{N}} I(u) \geq \inf_{u\in\mathcal{N}} \left(\frac{1}{2} - \frac{1}{\mu}\right)\|u\|^2 \geq \left(\frac{1}{2} - \frac{1}{\mu}\right)\left(\frac{1}{2C_2}\right)^{\frac{2}{p-2}} > 0,$$

by Lemma 2.5.5.   □

Lemma 2.5.7 *There exists* $u \in \mathcal{N}$ *such that* $I(u) = m$ *and* $u \geq 0$ *a.e. in* Ω.

Proof Let $\{u_k\}_k \subset \mathcal{N}$ be a minimizing sequence for I. The fact that $F(t)$ and $f(t)t$ are even implies that also $\{|u_k|\}_k$ is a minimizing sequence, so that we can assume from the beginning that $u_k(x) \geq 0$ a.e. in Ω. Since I is coercive on $\mathcal{N}$, the sequence $\{u_k\}_k$ is bounded and then, up to subsequences,

$$u_k \rightharpoonup u \quad \text{in } H^1_0(\Omega), \qquad u_k \to u \quad \text{in } L^q(\Omega)\; \forall q \in [2, 2^*),$$
$$u_k(x) \to u(x) \quad \text{a.e. in } \Omega.$$

This shows that $u(x) \geq 0$ a.e. and, with arguments already used many times,

$$\int_\Omega F(u_k)\,dx \to \int_\Omega F(u)\,dx,$$

$$\int_\Omega f(u_k)u_k\,dx \to \int_\Omega f(u)u\,dx, \quad \|u\|^2 \leq \liminf_{k\to+\infty} \|u_k\|^2.$$

Clearly the assumption $p < 2^*$ is essential in the first two limits.

Thus we deduce that

$$I(u) \leq \liminf_k I(u_k) = m \quad \text{and} \quad \|u\|^2 \leq \int_\Omega f(u)u\,dx.$$

Now we notice that from $\|u_k\|^2 = \int_\Omega f(u_k)u_k\,dx$ we have

$$\int_\Omega f(u)u\,dx = \lim_k \|u_k\|^2.$$

Since

$$\|u_k\|^2 \geq \left(\frac{1}{2C_2}\right)^{\frac{2}{p-2}} > 0$$

for all k we obtain

$$\int_\Omega f(u)u\,dx > 0,$$

which shows that it cannot be $u \equiv 0$.

If $\|u\|^2 = \int_\Omega f(u)u\,dx$, then $u \in \mathcal{N}$, u is the required minimum and the proof is complete. It remains to show that

$$\|u\|^2 < \int_\Omega f(u)u\,dx \tag{2.42}$$

leads to a contradiction.

To this aim, assume that (2.42) holds and consider the function

$$\gamma(t) = t^2\|u\|^2 - \int_\Omega f(tu)tu\,dx.$$

We already know that $\gamma(t) > 0$ in a right neighborhood of 0. The inequality (2.42) tells us that $\gamma(1) < 0$. Then there exists $t^* \in (0,1)$ such that $\gamma(t^*) = 0$, which means that $t^*u \in \mathcal{N}$. We then obtain

$$m \leq I(t^*u) = \int_\Omega \left[\frac{1}{2}f(t^*u)t^*u - F(t^*u)\right]dx.$$

From ($\mathbf{f_4}$) it is easy to see that the function

$$s \to \frac{1}{2}f(s)s - F(s)$$

is strictly increasing in $(0, +\infty)$, and then, since $u \geq 0$ and $t^* \in (0,1)$, we see that

$$m \leq \int_\Omega \left[\frac{1}{2}f(t^*u)t^*u - F(t^*u)\right]dx < \int_\Omega \left[\frac{1}{2}f(u)u - F(u)\right]dx$$

$$= \lim_k \int_\Omega \left[\frac{1}{2}f(u_k)u_k - F(u_k)\right]dx = \lim_k I(u_k) = m.$$

This contradiction shows that (2.42) cannot hold and concludes the proof.   □

The last step consists in showing that u is a critical point for I.

Lemma 2.5.8 *Let $u \in \mathcal{N}$ be such that $I(u) = m$. Then $I'(u) = 0$.*

Proof For every $v \in H_0^1(\Omega)$ there exists $\varepsilon > 0$ such that $u + sv \not\equiv 0$ for all s in $(-\varepsilon, \varepsilon)$. We know that there exists $t(s) \in \mathbb{R}$ such that $t(s)(u + sv) \in \mathcal{N}$. We now deduce some properties of the function $t(s)$. To this aim, consider the function

$$\varphi(s,t) = t^2\|u + sv\|^2 - \int_\Omega f(t(u + sv))t(u + sv)\,dx,$$

defined for $(s, t) \in (-\varepsilon, \varepsilon) \times \mathbb{R}$. Since $u \in \mathcal{N}$, we have

$$\varphi(0, 1) = \|u\|^2 - \int_\Omega f(u)u \, dx = 0.$$

Moreover, by $(\mathbf{f_2})$ and $(\mathbf{f_4})$, φ is a C^1 function and

$$\frac{\partial \varphi}{\partial t}(0, 1) = 2\|u\|^2 - \int_\Omega f(u)u \, dx - \int_\Omega f'(u)u^2 \, dx$$

$$= \int_\Omega \left[f(u)u - f'(u)u^2 \right] dx < 0.$$

Then, by the Implicit Function Theorem, for ε small there exists a C^1 function $t(s)$ such that $\varphi(s, t(s)) = 0$ and $t(0) = 1$. This also shows that $t(s) \neq 0$, at least for ε very small. Therefore $t(s)(u + sv) \in \mathcal{N}$.

Then setting

$$\gamma(s) = I(t(s)(u + sv)),$$

we have that γ is differentiable and has a minimum at $s = 0$; therefore

$$0 = \gamma'(0) = I'(t(0)u)(t'(0)u + t(0)v) = t'(0)I'(u)u + I'(u)v = I'(u)v.$$

Since v is arbitrary in $H_0^1(\Omega)$, we deduce that $I'(u) = 0$. $\qquad\square$

Example 2.5.9 We list a couple of examples of nonlinearities that satisfy assumptions $(\mathbf{f_1})$–$(\mathbf{f_4})$. These are functions that behave "like powers" without being exact powers. Since we want odd functions, we define them only for $t > 0$, and we extend them by oddness.

- $f(t) = t^{p-1} + t^{q-1}$, with $p, q \in (2, 2^*)$ and $p \neq q$.
- Slightly more generally, $f(t) = \sum_{i=1}^M a_i t^{p_i - 1}$ with $a_i > 0$ and $p_i \in (2, 2^*)$.
- $f(t) = \frac{t^{q-1}}{1+t^{q-p}}$, with $p \in (2, 2^*)$ and $q > 2$.

Remark 2.5.10 Let $2 < q < p < 2^*$; the function

$$f(t) = t^{p-1} - t^{q-1} \tag{2.43}$$

for $t > 0$ and extended by oddness to $\mathbb{R}$, does not satisfy the assumptions $(\mathbf{f_3})$ and $(\mathbf{f_4})$. However if one repeats the arguments carried out in the previous lemmas, one see that they still work, with minor changes. Therefore Problem (2.39) admits a solution also for this particular f. We now try to generalize this remark to a wider class of nonlinearities that contains (2.43).

We consider the problem

$$\begin{cases} -\Delta u + q(x)u = f(u) - g(u) & \text{in } \Omega, \\ u = 0 & \text{on } \partial\Omega. \end{cases} \tag{2.44}$$

We set $G(t) = \int_0^t g(s) \, ds$ and we introduce the following assumptions. Here $\mu > 2$, as in $(\mathbf{f_3})$.

(g₁) $g : \mathbb{R} \to \mathbb{R}$ is of class C^1 and is odd;

(g₂) $g'(0) = 0$ and there exists $\theta \in (2, \mu)$ such that $\limsup_{t \to +\infty} \frac{|g'(t)|}{t^{\theta - 2}} < +\infty$;

(g₃) the inequalities $g(t) \geq 0$ and $g'(t)t \leq (\mu - 1)g(t)$ hold for every $t \in (0, +\infty)$.

As far as f is concerned, we have to replace **(f₃)** and **(f₄)** with the stronger assumption

(f₅) the inequality $f'(t)t \geq (\mu - 1)f(t) > 0$ holds for every $t \in (0, +\infty)$.

We can now prove the following result.

Theorem 2.5.11 *Assume that* **(h₁)**, **(f₁)**, **(f₂)**, **(f₅)** *and* **(g₁)**–**(g₃)** *hold. Then Problem* (2.44) *admits at least one nontrivial and nonnegative solution.*

The proof follows exactly the same lines as the previous one. We start by noticing that since **(f₅)** implies **(f₃)** and **(f₄)**, all the properties of the functions $F(t)$ and $f(t)t$ obtained earlier are still valid. Concerning g and G, we list the properties that can be obtained working as in Lemma 2.5.3.

Lemma 2.5.12 *We have*

(1) $\lim_{t \to 0} \frac{G(t)}{t^2} = 0$.

(2) *There exist positive constants* M_5, M_6 *such that*

$$|g'(t)| \leq M_5 |t|^{\theta - 2} \quad \text{for } |t| > M_6.$$

(3) *There exist positive constants* D_1, D_2 *such that*

$$|G(t)| + |g(t)t| \leq D_1 t^2 + D_2 |t|^\theta \quad \forall t \in \mathbb{R}. \tag{2.45}$$

Solutions of Problem (2.44) are critical points of the functional

$$I(u) = \frac{1}{2} \|u\|^2 - \int_\Omega F(u)\,dx + \int_\Omega G(u)\,dx$$

and the corresponding Nehari manifold is

$$\mathcal{N} = \{u \in H_0^1(\Omega) \mid I'(u)u = 0, u \neq 0\}$$
$$= \left\{ u \in H_0^1(\Omega) \;\middle|\; \|u\|^2 = \int_\Omega f(u)u\,dx - \int_\Omega g(u)u\,dx, u \neq 0 \right\}.$$

The first step consist of course in checking that we can work in $\mathcal{N}$.

Lemma 2.5.13 *The Nehari manifold* $\mathcal{N}$ *is not empty.*

Proof Repeating the argument of Lemma 2.5.4, we fix $u \in H_0^1(\Omega)$ with $u \neq 0$ and $u \geq 0$, and we consider the function

$$\gamma(t) = I'(tu)tu = t^2 \|u\|^2 - \int_\Omega f(tu)tu\,dx + \int_\Omega g(tu)tu\,dx.$$

We already know that

$$\int_\Omega f(tu)tu\,dx = o(t^2)$$

as $t \to 0$, and the same ideas show that also

$$\int_\Omega g(tu)tu\,dx = o(t^2),$$

so that

$$\gamma(t) = t^2\|u\|^2 + o(t^2)$$

as $t \to 0$.

On the other hand, from the properties of g we deduce that for a suitable $C > 0$,

$$\left| \int_\Omega g(tu)tu\,dx \right| \leq Ct^2 + Ct^\theta = o(t^\mu),$$

when $t \to +\infty$. From this and the proof of Lemma 2.5.4 we immediately obtain

$$\gamma(t) \leq t^2\|u\|^2 - Dt^\mu \int_\Omega u^\mu\,dx + o(t^\mu)$$

as $t \to +\infty$. This proves the existence of $t > 0$ such that $tu \in \mathcal{N}$. We conclude the proof as in Lemma 2.5.4 when u has arbitrary sign. □

Defining as usual

$$m = \inf_{u \in \mathcal{N}} I(u),$$

we proceed to prove that m is attained by a function that solves Problem (2.44).

Lemma 2.5.14 *There results*

$$\inf_{u \in \mathcal{N}} \|u\|^2 > 0.$$

Proof Working as in Lemma 2.5.5 and using the fact that $g(u)u \geq 0$, we obtain for every $u \in \mathcal{N}$,

$$\|u\|^2 = \int_\Omega f(u)u\,dx - \int_\Omega g(u)u\,dx \leq \frac{1}{2}\lambda_1 \int_\Omega u^2\,dx + C_1 \int_\Omega |u|^p\,dx$$

$$\leq \frac{1}{2}\|u\|^2 + C_2\|u\|^p,$$

which allows us to conclude exactly as above. □

Lemma 2.5.15 *The functional I is coercive on $\mathcal{N}$ and $m > 0$.*

Proof From (**g3**) we deduce that $\mu G(u) \geq g(u)u$. Then, if $u \in \mathcal{N}$, we have

$$I(u) = \frac{1}{2}\|u\|^2 - \int_\Omega F(u)\,dx + \int_\Omega G(u)\,dx$$

$$= \left(\frac{1}{2} - \frac{1}{\mu}\right)\|u\|^2 + \int_\Omega \left(\frac{1}{\mu}f(u)u - F(u)\right)dx$$

$$+ \int_\Omega \left(G(u) - \frac{1}{\mu}g(u)u\right)dx \geq \left(\frac{1}{2} - \frac{1}{\mu}\right)\|u\|^2,$$

and we conclude as in Lemma 2.5.6. $\qquad\square$

We can now prove that the infimum of I on $\mathcal{N}$ is attained.

Lemma 2.5.16 *There exists* $u \in \mathcal{N}$ *such that* $I(u) = m$ *and* $u \geq 0$.

Proof Let $\{u_k\}_k \subset \mathcal{N}$ be a minimizing sequence for I. As above, we can assume that $u_k \geq 0$. By Lemma 2.5.15, I is coercive on $\mathcal{N}$, and hence $\{u_k\}_k$ is bounded. Up to subsequences,

$$u_k \rightharpoonup u \quad \text{in } H_0^1(\Omega), \qquad u_k \to u \quad \text{in } L^q(\Omega) \; \forall q \in [2, 2^*),$$

$$u_k(x) \to u(x) \quad \text{a.e. in } \Omega.$$

Then $u(x) \geq 0$, and with the usual arguments, as $k \to \infty$,

$$\int_\Omega F(u_k)\,dx \to \int_\Omega F(u)\,dx, \qquad \int_\Omega f(u_k)u_k\,dx \to \int_\Omega f(u)u\,dx,$$

$$\int_\Omega G(u_k)\,dx \to \int_\Omega G(u)\,dx,$$

$$\int_\Omega g(u_k)u_k\,dx \to \int_\Omega g(u)u\,dx, \qquad \|u\|^2 \leq \liminf_{k\to+\infty}\|u_k\|^2.$$

These yield immediately

$$I(u) \leq \liminf_k I(u_k) = m \quad \text{and} \quad \|u\|^2 \leq \int_\Omega f(u)u\,dx - \int_\Omega g(u)u\,dx.$$

Since

$$\int_\Omega f(u)u\,dx - \int_\Omega g(u)u\,dx = \lim_k \|u_k\|^2 \geq \left(\frac{1}{2C_2}\right)^{\frac{2}{p-2}} > 0,$$

we obtain

$$\int_\Omega f(u)u\,dx - \int_\Omega g(u)u\,dx > 0,$$

which shows that it cannot be $u \equiv 0$.

Now, if

$$\|u\|^2 = \int_\Omega f(u)u\,dx - \int_\Omega g(u)u\,dx,$$

then $u \in \mathcal{N}$, and u is the required minimum. We must therefore show that it cannot be

$$\|u\|^2 < \int_\Omega f(u)u\,dx - \int_\Omega g(u)u\,dx. \tag{2.46}$$

To this aim, we consider the function

$$\gamma(t) = t^2\|u\|^2 - \int_\Omega f(tu)tu\,dx + \int_\Omega g(tu)tu\,dx.$$

We have already noticed that $\gamma(t) > 0$ in a right neighborhood of zero. Assuming (2.46) means that $\gamma(1) < 0$, hence there exists $t^* \in (0, 1)$ such that $\gamma(t^*) = 0$, namely $t^*u \in \mathcal{N}$. Then we see that

$$m \le I(t^*u) = \frac{1}{2}\|t^*u\|^2 - \int_\Omega F(t^*u)\,dx + \int_\Omega G(t^*u)\,dx$$

$$= \left(\frac{1}{2} - \frac{1}{\mu}\right)\|t^*u\|^2 + \int_\Omega \left(\frac{1}{\mu}f(t^*u)t^*u - F(t^*u)\right)dx$$

$$+ \int_\Omega \left(G(t^*u) - \frac{1}{\mu}g(t^*u)t^*u\right)dx.$$

From assumptions (**f₅**) and (**g₃**) we deduce immediately that the functions

$$s \mapsto \frac{1}{\mu}f(s)s - F(s) \quad \text{and} \quad s \mapsto G(s) - \frac{1}{\mu}g(s)s$$

are nondecreasing in $(0, +\infty)$, so that being $u \ge 0$ and $t^* \in (0, 1)$, we have

$$m \le \left(\frac{1}{2} - \frac{1}{\mu}\right)\|t^*u\|^2 + \int_\Omega \left(\frac{1}{\mu}f(t^*u)t^*u - F(t^*u)\right)dx$$

$$+ \int_\Omega \left(\frac{1}{\mu}G(t^*u) - g(t^*u)t^*u\right)dx$$

$$< \left(\frac{1}{2} - \frac{1}{\mu}\right)\|u\|^2 + \int_\Omega \left(\frac{1}{\mu}f(u)u - F(u)\right)dx + \int_\Omega \left(\frac{1}{\mu}G(u) - g(u)u\right)dx$$

$$\le \liminf_k \left(\frac{1}{2} - \frac{1}{\mu}\right)\|u_k\|^2 + \lim_k \int_\Omega \left[\frac{1}{\mu}f(u_k)u_k - F(u_k)\right]dx$$

$$+ \lim_k \int_\Omega \left[G(u_k) - \frac{1}{\mu}g(u_k)u_k\right]dx$$

$$= \liminf_k I(u_k) = m.$$

This contradiction shows that (2.46) cannot hold, and the proof is complete. □

Lastly we prove that the minimizer u is a critical point of I on $H_0^1(\Omega)$.

Lemma 2.5.17 *Let $u \in \mathcal{N}$ be such that $I(u) = m$. Then $I'(u) = 0$.*

Proof The technique is the same as in Lemma 2.5.13. Take $v \in H_0^1(\Omega)$ and let $\varepsilon > 0$ be so small that $u + sv \neq 0$ for all $s \in (-\varepsilon, \varepsilon)$. We know that there exists $t(s) \in \mathbb{R}$ such that $t(s)(u + sv) \in \mathcal{N}$. Consider the function

$$\varphi(s, t) = t^2 \|u + sv\|^2 - \int_\Omega f(t(u + sv))t(u + sv)\,dx$$
$$+ \int_\Omega g(t(u + sv))t(u + sv)\,dx,$$

defined for $(s, t) \in (-\varepsilon, \varepsilon) \times \mathbb{R}$. Since $u \in \mathcal{N}$, we have

$$\varphi(0, 1) = \|u\|^2 - \int_\Omega f(u)u\,dx + \int_\Omega g(u)u\,dx = 0.$$

On the other hand, by $(\mathbf{f_5})$ and $(\mathbf{g_3})$,

$$\frac{\partial \varphi}{\partial t}(0, 1) = 2\|u\|^2 - \int_\Omega f(u)u\,dx - \int_\Omega f'(u)u^2\,dx + \int_\Omega g(u)u\,dx$$
$$+ \int_\Omega g'(u)u^2\,dx$$
$$\leq 2\|u\|^2 - \int_\Omega f(u)u\,dx - (\mu - 1)\int_\Omega f(u)u\,dx + \int_\Omega g(u)u\,dx$$
$$+ (\mu - 1)\int_\Omega g(u)u\,dx$$
$$= 2\|u\|^2 - \mu \int_\Omega f(u)u\,dx + \mu \int_\Omega g(u)u\,dx = (2 - \mu)\|u\|^2 < 0.$$

So, by the Implicit Function Theorem, for ε small enough we can determine a function $t \in C^1(-\varepsilon, \varepsilon)$ such that $f(s, t(s)) = 0$ and $t(0) = 1$. This says that $t(s) \neq 0$, at least for ε very small, and then $t(s)(u + sv) \in \mathcal{N}$.

Setting

$$\gamma(s) = I(t(s)(u + sv)),$$

we obtain that γ is differentiable and has a minimum point at $s = 0$; thus

$$0 = \gamma'(0) = I'(t(0)u)(t'(0)u + t(0)v) = t'(0)I'(u)u + I'(u)v = I'(u)v.$$

Since $v \in H_0^1(\Omega)$, is arbitrary, we conclude that $I'(u) = 0$. $\qquad\square$

Example 2.5.18 Also in this case it is easy to produce examples of nonlinearities that satisfy the assumption of Theorem 2.5.11. As usual we define the functions for $t > 0$ and we extend them by oddness.

- $f(t) = t^{p-1}$, $g(t) = t^{q-1}$, where $2 < q < p < 2^*$.

  Slightly more generally, $f(t) = \sum_{i=1}^M a_i t^{p_i-1}$ and $g(t) = \sum_{i=1}^N b_i t^{q_i-1}$ with $a_i, b_i > 0$, $p_i, q_i \in (2, 2^*)$ and $\min\{p_i\} > \max\{q_i\}$.
- $f(t) = \frac{s^{q-1}}{1+s^{q-p}}$, where $p \in (2, 2^*)$, $q > 2$ and g as in the preceding example, with $\max\{q_i\} < \min\{p, q\}$.

2.6 The p-Laplacian

In this section we leave temporarily the semilinear equations studied up to now to give an example of how the techniques introduced so far can be successfully applied, and with almost no changes, to more general *quasilinear* equations. Although quasilinear equations are in general much more difficult than semilinear ones, a certain number of results can be proved by the application of variational methods with surprisingly little effort.

This section should be considered as an *example* and can be skipped on first reading.

The particular type of equations we consider are the so-called p-Laplace equations, that are commonly seen as the simplest generalizations of the Laplacian to the quasilinear context.

The p-Laplacian is the second order non linear differential operator defined as

$$\Delta_p u = \operatorname{div}\big(|\nabla u|^{p-2}\nabla u\big).$$

Here $p > 1$, and of course when $p = 2$ the p-Laplacian is the usual Laplacian.

Notice that this operator is *linear* in the second derivatives, but contains terms which are *nonlinear* functions of the *gradient* of the unknown.

When u is regular enough an easy computation shows that

$$\Delta_p u = |\nabla u|^{p-2}\Delta u + (p-2)|\nabla u|^{p-4}\sum_{i,j=1}^{n}\frac{\partial^2 u}{\partial x_i \partial x_j}\frac{\partial u}{\partial x_i}\frac{\partial u}{\partial x_j},$$

a rather complicated expression. For this reason, it is much more convenient to use some weak formulation for this operator. To write down such a weak form we need, as in the previous chapters, to introduce suitable function spaces. We will list a number of definitions and results that generalize those we have seen in Chap. 1 for the Sobolev spaces $H^1(\Omega)$ and $H_0^1(\Omega)$.

- $W^{1,p}(\Omega)$, for $p \in [1,+\infty)$, is the Sobolev space defined by

$$W^{1,p}(\Omega) = \left\{ u \in L^p(\Omega) \;\middle|\; \frac{\partial u}{\partial x_i} \in L^p(\Omega), i = 1, \dots, N \right\},$$

where the derivative $\frac{\partial u}{\partial x_i}$ is in the sense of distributions. The spaces $W^{1,p}(\Omega)$ are reflexive Banach spaces when endowed with the norm

$$\|u\|_p = |u|_p + \sum_{i=1}^{N}\left|\frac{\partial u}{\partial x_i}\right|_p.$$

Recall that $|\cdot|_p$ is the L^p norm.
- $W_0^{1,p}(\Omega)$ is the closure of $C_0^\infty(\Omega)$ in $W^{1,p}(\Omega)$.
- If $u \in W_0^{1,p}(\Omega)$, then $|u|, u^+, u^- \in W_0^{1,p}(\Omega)$ and

$$\int_\Omega |\nabla|u||^p \, dx = \int_\Omega |\nabla u|^p \, dx.$$

We also recall, as in the previous chapters, some embedding theorems.

Theorem 2.6.1 *Let* $\Omega \subset \mathbb{R}^N$ *be open, bounded and have smooth boundary. Let* $p \geq 1$.

- *If* $p < N$, *then* $W^{1,p}(\Omega) \hookrightarrow L^q(\Omega)$ *for every* $q \in [1, \frac{Np}{N-p}]$; *the embedding is compact for every* $q \in [1, \frac{Np}{N-p})$.
- *If* $p = N$, *then* $W^{1,p}(\Omega) \hookrightarrow L^q(\Omega)$ *for every* $q \in [1, +\infty)$; *the embedding is compact.*
- *If* $p > N$, *then* $W^{1,p}(\Omega) \hookrightarrow L^\infty(\Omega)$; *the embedding is compact.*

The following particular case is the most used in our context, and we state it separately.

Theorem 2.6.2 *Let* Ω *be an open and bounded subset or* $\mathbb{R}^N$, *with* $N \geq 3$. *Let* $1 < p < N$. *Then*

$$W_0^{1,p}(\Omega) \hookrightarrow L^q(\Omega) \quad \text{for every } q \in \left[1, \frac{Np}{N-p}\right].$$

The embedding is compact for every $q \in [1, \frac{Np}{N-p})$.

The number $\frac{Np}{N-p}$ is denoted by p^* and is called the *critical Sobolev exponent* for the embedding of $W_0^{1,p}(\Omega)$ into L^q.

Theorem 2.6.3 (Poincaré inequality for $W_0^{1,p}$) *Let* $\Omega \subset \mathbb{R}^N$ *be open and bounded. Then there exists a constant* $C > 0$, *depending only on* Ω, *such that*

$$\int_\Omega |u|^p \, dx \leq C \int_\Omega |\nabla u|^p \, dx \quad \forall u \in W_0^{1,p}(\Omega).$$

Therefore, if Ω is a bounded open set, $(\int_\Omega |\nabla u|^p \, dx)^{\frac{1}{p}}$ is a norm on $W_0^{1,p}(\Omega)$, equivalent to the standard one.

2.6.1 Basic Theory

We begin by studying the differentiability of some relevant functionals on $W_0^{1,p}(\Omega)$. We start with the norm.

Theorem 2.6.4 *Let* $\Omega \subseteq \mathbb{R}^N$, $N \geq 3$, *be an open set. For* $p \in (1, +\infty)$, *define a functional* $J : W_0^{1,p}(\Omega) \to \mathbb{R}$ *by*

$$J(u) = \int_\Omega |\nabla u|^p \, dx.$$

Then J is differentiable in $W_0^{1,p}(\Omega)$ and

$$J'(u)v = p \int_\Omega |\nabla u|^{p-2} \nabla u \cdot \nabla v \, dx.$$

Proof Consider the function $\varphi : \mathbb{R}^N \to \mathbb{R}$, defined by $\varphi(x) = |x|^p$. It is a C^1 function and $\nabla \varphi(x) = p|x|^{p-2}x$, so that, for all $x, y \in \mathbb{R}^N$,

$$\lim_{t \to 0} \frac{\varphi(x + ty) - \varphi(x)}{t} = p|x|^{p-2}x \cdot y.$$

As a consequence,

$$\lim_{t \to 0} \frac{|\nabla u(x) + t\nabla v(x)|^p - |\nabla u(x)|^p}{t} = p|\nabla u(x)|^{p-2}\nabla u(x) \cdot \nabla v(x) \quad \text{a.e. in } \Omega.$$

By the Lagrange Theorem there exists $\theta \in \mathbb{R}$ such that $|\theta| \le |t|$ and

$$\left| \frac{|\nabla u + t\nabla v|^p - |\nabla u|^p}{t} \right| \le p\left| |\nabla u + \theta\nabla v|^{p-2}(\nabla u + \theta\nabla v) \cdot \nabla v \right|$$

$$\le C\left(|\nabla u|^{p-1}|\nabla v| + |\nabla v|^p \right) \in L^1(\Omega).$$

By dominated convergence we obtain

$$\lim_{t \to 0} \int_\Omega \frac{|\nabla u + t\nabla v|^p - |\nabla u|^p}{t} \, dx = p \int_\Omega |\nabla u|^{p-2}\nabla u \cdot \nabla v \, dx,$$

so that J is Gâteaux differentiable and

$$J_G'(u)v = p \int_\Omega |\nabla u|^{p-2}\nabla u \cdot \nabla v \, dx.$$

We have now to prove that $J_G' : W_0^{1,p}(\Omega) \to [W^{1,p}(\Omega)]'$ is continuous. To this aim we take a sequence $\{u_k\}_k$ in $W_0^{1,p}(\Omega)$ such that $u_k \to u$ in $W_0^{1,p}(\Omega)$. In particular we can assume, as usual, that up to subsequences,

- $\nabla u_k \to \nabla u$ in $(L^p(\Omega))^N$ as $k \to \infty$;
- $\nabla u_k(x) \to \nabla u(x)$ a.e. as $k \to \infty$;
- there exists $w \in L^1(\Omega)$ such that $|\nabla u_k(x)|^p \le w(x)$ a.e. in Ω, and for all $k \in \mathbb{N}$.

We have

$$(J_G'(u) - J_G'(u_k))v = p \int_\Omega (|\nabla u|^{p-2}\nabla u - |\nabla u_k|^{p-2}\nabla u_k) \cdot \nabla v \, dx$$

and

$$\left| \int_\Omega (|\nabla u|^{p-2}\nabla u - |\nabla u_k|^{p-2}\nabla u_k) \cdot \nabla v \, dx \right|$$

$$\le \left(\int_\Omega \left| |\nabla u_k|^{p-2}\nabla u_k - |\nabla u|^{p-2}\nabla u \right|^{\frac{p}{p-1}} dx \right)^{\frac{p-1}{p}} \left(\int_\Omega |\nabla v|^p \, dx \right)^{1/p}$$

$$\le \left(\int_\Omega \left| |\nabla u_k|^{p-2}\nabla u_k - |\nabla u|^{p-2}\nabla u \right|^{\frac{p}{p-1}} dx \right)^{\frac{p-1}{p}} \|v\|,$$

so that

$$\|J_G'(u) - J_G'(u_k)\| = \sup\left\{\left|(J_G'(u) - J_G'(u_k))v\right| \mid v \in W^{1,p}(\Omega), \|v\| = 1\right\}$$

$$\leq \left(\int_\Omega \left||\nabla u_k|^{p-2}\nabla u_k - |\nabla u|^{p-2}\nabla u\right|^{\frac{p}{p-1}} dx\right)^{\frac{p-1}{p}}.$$

Now we know that

$$|\nabla u_k(x)|^{p-2}\nabla u_k(x) \to |\nabla u(x)|^{p-2}\nabla u(x)$$

almost everywhere in Ω, and that

$$\left||\nabla u_k|^{p-2}\nabla u_k - |\nabla u|^{p-2}\nabla u\right|^{\frac{p}{p-1}} \leq C(|\nabla u_k|^p + |\nabla u|^p) \leq w + |\nabla u|^p \in L^1(\Omega).$$

By dominated convergence we then obtain

$$\int_\Omega \left||\nabla u_k|^{p-2}\nabla u_k - |\nabla u|^{p-2}\nabla u\right|^{\frac{p}{p-1}} dx \to 0,$$

and hence $\|J_G'(u_k) - J_G'(u)\| \to 0$. This holds for a subsequence of the original sequence $\{u_k\}$, but we can argue as before to obtain that J_G' is a continuous function, so that J is differentiable with differential given by

$$J'(u)v = p\int_\Omega |\nabla u|^{p-2}\nabla u \cdot \nabla v\, dx. \qquad \square$$

Remark 2.6.5 When $p \neq 2$, $W_0^{1,p}(\Omega)$ is *not* a Hilbert space. So here is an example of a Banach space where the norm is differentiable, see Remark 1.3.15.

Theorem 2.6.6 *Let $\Omega \subset \mathbb{R}^N$, $N \geq 3$, be a bounded open set. Take a number $1 < p < N$ and let $p^* = \frac{Np}{N-p}$ be the critical exponent for the embedding of $W_0^{1,p}(\Omega)$ in $L^q(\Omega)$. Let $f : \mathbb{R} \to \mathbb{R}$ be a continuous function, and assume that there exist $a, b > 0$ such that*

$$|f(t)| \leq a + b|t|^{p^*-1} \tag{2.47}$$

for all $t \in \mathbb{R}$. Define

$$F(t) = \int_0^t f(s)\, ds$$

and consider the functional $J : W_0^{1,p}(\Omega) \to \mathbb{R}$ given by

$$J(u) = \int_\Omega F(u(x))\, dx.$$

Then J is differentiable on $W_0^{1,p}(\Omega)$ and

$$J'(u)v = \int_\Omega f(u(x))v(x)\, dx \quad \text{for all } u, v \in W_0^{1,p}(\Omega).$$

We do not write down the proof of this result, because the argument is very similar to the one of Example 1.3.20.

We can now state a definition of weak solution for equations involving the p-Laplacian. The idea for a weak formulation can be obtained, as in the usual case $p = 2$, using integration by parts. Let us assume that u and v are functions defined on a bounded open set Ω, that they vanish on $\partial\Omega$, and are smooth enough for the following computations to make sense. Integrating by parts we obtain

$$\int_\Omega v \, \Delta_p u \, dx = \int_\Omega v \, \mathrm{div}\big(|\nabla u|^{p-2}\nabla u\big) \, dx = -\int_\Omega |\nabla u|^{p-2}\nabla u \cdot \nabla v \, dx.$$

This suggests that a good candidate as a weak form for the p-Laplacian is the operator

$$v \mapsto -\int_\Omega |\nabla u|^{p-2}\nabla u \cdot \nabla v \, dx.$$

We now give a precise formulation to the preceding heuristic discussion, at least for the cases that we want to treat. Let us consider $1 < p < N$ and take $h \in L^{p'}$, where $p' = \frac{p}{p-1}$ is the conjugate exponent of p. Let $f : \mathbb{R} \to \mathbb{R}$ be a continuous function satisfying the growth assumption (2.47). We will say that u is a weak solution of the boundary value problem

$$\begin{cases} -\Delta_p u = f(u) + h & \text{in } \Omega, \\ u = 0 & \text{on } \partial\Omega \end{cases} \tag{2.48}$$

if $u \in W_0^{1,p}(\Omega)$ and

$$\int_\Omega |\nabla u|^{p-2}\nabla u \cdot \nabla v \, dx = \int_\Omega f(u)v \, dx + \int_\Omega hv \, dx \quad \forall v \in W_0^{1,p}(\Omega).$$

Notice that, as in the case of the Laplacian, the boundary conditions are englobed in the function space $W_0^{1,p}(\Omega)$. Thanks to the previous results about differentiation of functionals, it is easy to spot the link between weak solutions and critical points. Indeed, denoting $F(t) = \int_0^t f(s) \, ds$, we can define a functional $J : W_0^{1,p}(\Omega) \to \mathbb{R}$ by

$$J(u) = \int_\Omega |\nabla u|^p \, dx - \int_\Omega F(u) \, dx - \int_\Omega hu \, dx.$$

We know that J is differentiable on $W_0^{1,p}(\Omega)$, with

$$J'(u)v = \int_\Omega |\nabla u|^{p-2}\nabla u \cdot \nabla v \, dx - \int_\Omega f(u)v \, dx - \int_\Omega hv \, dx.$$

Hence, a critical point of J is exactly a weak solution of (2.48) in the sense previously stated.

Remark 2.6.7 The procedure to obtain a classical solution from a weak one is more problematic in the case of quasilinear equations. Indeed, a weak solution which is regular enough is also a classical solution, but the regularity results for weak solutions of p-Laplacian equations that have been obtained so far are not completely satisfactory: weak solution may be not regular enough to be also classical solutions.

We conclude with a simple existence result for equations with right-hand-side independent of u.

Theorem 2.6.8 *Let $\Omega \subset \mathbb{R}^N$ be a bounded open set and let $p > 1$. Let $p' = \frac{p}{p-1}$ be the conjugate exponent of p. Then, for every $h \in L^{p'}(\Omega)$, the problem*

$$\begin{cases} -\Delta_p u = h & \text{in } \Omega, \\ u = 0 & \text{on } \partial\Omega \end{cases} \tag{2.49}$$

has a unique (weak) solution.

Proof Define a functional $J : W_0^{1,p}(\Omega) \to \mathbb{R}$ by

$$J(u) = \frac{1}{p} \int_\Omega |\nabla u|^p \, dx - \int_\Omega h u \, dx.$$

We know that J is differentiable on $W_0^{1,p}(\Omega)$ with

$$J'(u)v = \int_\Omega |\nabla u|^{p-2} \nabla u \cdot \nabla v \, dx - \int_\Omega h v \, dx.$$

It is well known that the function $x \to |x|^p$, $x \in \mathbb{R}^N$, is strictly convex. From this, it is easy to deduce that J is strictly convex. Moreover,

$$J(u) \geq \frac{1}{p} \|u\|^p - |h|_{p'} |u|_p \geq \frac{1}{p} \|u\|^p - C\|u\|,$$

so that J is coercive because $p > 1$. Applying Theorems 1.5.6 and 1.5.8, we find that J has a unique critical point which is the unique (weak) solution of Problem (2.49). $\qquad\square$

Remark 2.6.9 Notice that the previous theorem holds for any $p > 1$.

Remark 2.6.10 In Sect. 1.6 we have seen similar results. In that case we dealt with linear equations, and it is well known that they can be solved by different methods, for example by a straightforward application of the Riesz Theorem 1.2.9. On the contrary, (2.49), being quasilinear, cannot be treated by the standard methods of the linear theory, available for example for (1.22) and (1.23). Variational methods give an unified treatment for linear and nonlinear problems.

2.6.2 Two Applications

We start with a nonlinear eigenvalue problem. Indeed, it is possible to generalize part of the theory and of the results concerning linear eigenvalue problems to nonlinear problems. Nonlinear eigenvalue problems are a largely studied subject in recent research. For some references on these topics, start for example from [20].

Recall that in the Banach space $W_0^{1,p}(\Omega)$ we use the norm $\|u\|^p = \int_\Omega |\nabla u|^p \, dx$.

Theorem 2.6.11 *Let $\Omega \subset \mathbb{R}^N$, $N \geq 3$, be a bounded open set and let $1 < p < N$. Define*

$$\mu = \inf_{u \in W_0^{1,p}(\Omega) \setminus \{0\}} \frac{\int_\Omega |\nabla u|^p \, dx}{\int_\Omega |u|^p \, dx} = \inf_{u \in W_0^{1,p}(\Omega) \setminus \{0\}} \frac{\|u\|^p}{|u|_p^p}.$$

Then μ is positive and is achieved by some $u \in W_0^{1,p}(\Omega) \setminus \{0\}$. The function u can be chosen nonnegative and satisfies (weakly)

$$\begin{cases} -\operatorname{div}\left(|\nabla u|^{p-2}\nabla u\right) = \mu u^{p-1} & \text{in } \Omega, \\ u = 0 & \text{on } \partial\Omega. \end{cases} \qquad (2.50)$$

Proof For $u \in W_0^{1,p}(\Omega)$, set $I(u) = \|u\|^p$, $J(u) = |u|_p^p$ and define the quotient functional $Q : W_0^{1,p}(\Omega) \setminus \{0\} \to \mathbb{R}$ as

$$Q(u) = \frac{I(u)}{J(u)}.$$

Then

$$\mu = \inf_{u \in W_0^{1,p}(\Omega) \setminus \{0\}} Q(u).$$

The Poincaré inequality immediately gives $\mu > 0$. Let $\{u_k\}_k$ be a minimizing sequence. Clearly $|u_k|$ is also a minimizing sequence for Q, so we can assume that $u_k(x) \geq 0$ a.e. in Ω. As the functional Q is homogeneous of degree zero, i.e. $Q(\lambda u) = Q(u)$ for every $\lambda \in \mathbb{R}$, we can normalize u_k by setting $|u_k|_p = 1$ for every k. Then we see that $J(u_k)$, namely the $W^{1,p}$ norm of u_k, must be bounded independently of k.

By the Sobolev embeddings (Theorem 2.6.2), we deduce that, up to subsequences,

- $u_k \rightharpoonup u$ in $W^{1,p}(\Omega)$;
- $u_k \to u$ in $L^p(\Omega)$;
- $u_k(x) \to u(x)$ a.e. in Ω;

In particular, $J(u) = 1$ and $u(x) \geq 0$ a.e. Then, by weak lower semicontinuity of the norm,

$$Q(u) = I(u) \leq \liminf_k I(u_k) = \liminf_k Q(u_k) = \mu,$$

so $u \in W_0^{1,p}(\Omega) \setminus \{0\}$ and $Q(u) = \mu$.

We have seen in the preceding subsection that I and J are differentiable. Then so is Q, and

$$Q'(u)v = \frac{1}{J(u)}\left(I'(u)v - Q(u)J'(u)v\right).$$

Since u is a minimum point for Q, we obtain $Q'(u) = 0$ and therefore

$$I'(u)v = Q(u)J'(u)v = \mu J'(u)v$$

for every $v \in W_0^{1,p}(\Omega)$, which is nothing but the weak form of (2.50). $\qquad\square$

We give a last result, concerning existence of a minimum for a coercive functional involving the p-Laplacian. This result generalizes Theorem 2.1.5.

Theorem 2.6.12 *Let Ω be an open bounded subset of $\mathbb{R}^N$, $N \geq 3$, and let $1 < p < N$. Let $f : \mathbb{R} \to \mathbb{R}$ be a continuous function such that*

$$|f(t)| \leq a + b|t|^{q-1} \quad \forall t \in \mathbb{R},$$

where $a, b > 0$ and $1 \leq q < p$. Then, for all $h \in L^{p'}(\Omega)$, where $p' = \frac{p}{p-1}$, there exists a weak solution of the problem

$$\begin{cases} -\operatorname{div}(|\nabla u|^{p-2}\nabla u) = f(u) + h(x) & \text{in } \Omega, \\ u = 0 & \text{on } \partial\Omega. \end{cases} \tag{2.51}$$

Proof We consider the functional $I : W_0^{1,p}(\Omega) \to \mathbb{R}$ defined by

$$\begin{aligned}
I(u) &= \frac{1}{p}\int_\Omega |\nabla u|^p dx - \int_\Omega F(u)\,dx - \int_\Omega hu\,dx \\
&= \frac{1}{p}\|u\|^p - \int_\Omega F(u)\,dx - \int_\Omega hu\,dx,
\end{aligned}$$

where as usual $F(t) = \int_0^t f(s)\,ds$. The hypotheses on f immediately imply that $F(t) \leq a_1 + b_1|t|^q$, and therefore, by the Hölder and Sobolev inequalities,

$$I(u) \geq \frac{1}{p}\|u\|^p - C_1 - C_2|u|_q^q - C_3|h|_{p'}|u|_p \geq C\big(\|u\|^p - \|u\|^q - \|u\| - 1\big),$$

which shows that I is coercive because $q < p$. Thus, any minimizing sequence u_k for I is bounded in $W_0^{1,p}(\Omega)$. Then we can extract a subsequence, still denoted u_k, that converges weakly to some $u \in W_0^{1,p}(\Omega)$; by compactness of the Sobolev embedding, we can make sure that it also converges strongly in $L^q(\Omega)$. Weak lower semicontinuity of the norm and the usual arguments show that the weak limit u is a minimum point for I. This gives $I'(u) = 0$, which is the weak form of (2.51). □

2.7 Exercises

1. Let X be a complete normed space, and let $a : X \times X \to \mathbb{R}$ be a continuous bilinear form. Let $\phi : X \to \mathbb{R}$ be the associated quadratic form, $\phi(u) = a(u, u)$. Consider the following statements.
   (a) $\forall u \neq 0$, $\phi(u) > 0$.
   (b) $\lim_{\|u\| \to \infty} \phi(u) = +\infty$.
   (c) There exists $\alpha > 0$ such that for every $u \in X$, $\phi(u) \geq \alpha\|u\|^2$.
   Prove that if X has finite dimension, then (a), (b) and (c) are equivalent. Prove that if X is infinite dimensional, then (c) $\Longleftrightarrow$ (b) $\Longrightarrow$ (a), but (a) needs not imply (b) or (c).

2. Let A be an $N \times N$ symmetric matrix. Prove that A has at least one eigenvalue by maximizing the function $\phi : \mathbb{R}^N \to \mathbb{R}$ defined by $\phi(x) = Ax \cdot x$ constrained on the unit sphere of $\mathbb{R}^N$.

3. Let $f : \mathbb{R} \to \mathbb{R}$ be continuous and satisfy $f(t) = 0$ if $t \le 0$. Assume that

$$t \mapsto \frac{f(t)}{t} \text{ is strictly decreasing on } (0, +\infty)$$

and denote $\alpha = \lim_{t \to 0^+} \frac{f(t)}{t}$ and $\beta = \lim_{t \to +\infty} \frac{f(t)}{t}$.

Let $\Omega \subset \mathbb{R}^N$ be a bounded open set. Prove that *if* the problem

$$\begin{cases} -\Delta u = f(u) & \text{in } \Omega, \\ u > 0 & \text{in } \Omega, \\ u = 0 & \text{on } \partial\Omega \end{cases}$$

has a solution, then $\alpha > \lambda_1$ and $\beta < \lambda_1$. Hint: multiply the equation by φ_1.

4. Let $\Omega \subset \mathbb{R}^N$, with $N \ge 3$, be open and bounded, $h \in L^\infty(\Omega) \setminus \{0\}$ and $2 < p < 2^*$, and consider the problem

$$\begin{cases} -\Delta u = h(x)|u|^{p-2}u & \text{in } \Omega, \\ u = 0 & \text{on } \partial\Omega. \end{cases}$$

(a) Prove that if $h(x) \le 0$ a.e. in Ω, the problem has no nontrivial solutions.

(b) If $h(x) > 0$ on a set of positive measure, prove that the problem has at least one nontrivial solution. *Hint*: consider the set

$$M = \left\{ u \in H_0^1(\Omega) \;\middle|\; \int_\Omega h(x)|u|^p \, dx = 1 \right\},$$

prove that $M \ne \emptyset$ and argue as in Sect. 2.3.1.

5. Let $\Omega \subset \mathbb{R}^N$, with $N \ge 3$, be open and bounded, and let $a \in L^\infty(\Omega)$ satisfy $a(x) > 0$ a.e. in Ω. Take $p \in (2, 2^*)$. Assume that the differentiable functional $Q : H_0^1(\Omega) \setminus \{0\} \to \mathbb{R}$ defined by

$$Q(u) = \frac{\int_\Omega |\nabla u|^2 \, dx}{(\int_\Omega a(x)|u|^p \, dx)^{2/p}}$$

has a critical point v. Show that a multiple of v is a weak solution of the problem

$$\begin{cases} -\Delta u = a(x)|u|^{p-2}u & \text{in } \Omega, \\ u = 0 & \text{on } \partial\Omega. \end{cases}$$

Hint: recall Example 1.3.16.

6. Let $\Omega \subset \mathbb{R}^N$, with $N \ge 3$, be open and bounded, let $a, b \in C(\overline{\Omega}) \setminus \{0\}$ and let $2 < p < q < 2^*$. Assume that $b \ge 0$ in Ω and that for a.e. x, $a(x) \ne 0$ implies $b(x) \ne 0$. Consider the problem

$$\begin{cases} -\Delta u = a(x)|u|^{p-2}u + b(x)|u|^{q-2}u & \text{in } \Omega, \\ u = 0 & \text{on } \partial\Omega. \end{cases} \tag{2.52}$$

(a) Write down an energy functional I such that $I'(u) = 0$ if and only if u is a weak solution for this problem.

(b) Prove that the set

$$\mathcal{N} = \left\{ u \in H^1_0(\Omega) \ \Big| \ I'(u)u = 0, \ \int_\Omega b(x)|u|^q \, dx > 0 \right\}$$

is not empty.

(c) Prove that there exists a minimum point u for I on $\mathcal{N}$ such that $u \geq 0$.

(d) Prove that the minimum is a weak solution of (2.52).

7. Let $\Omega \subset \mathbb{R}^N$, with $N \geq 3$, be open and bounded, and take numbers r, p such that $1 < r < 2 < p < 2^*$. Consider the problem

$$\begin{cases} -\Delta u = |u|^{p-2}u - |u|^{r-2}u & \text{in } \Omega, \\ u = 0 & \text{on } \partial\Omega. \end{cases}$$

Define the energy functional and the Nehari manifold for this problem, and prove the existence of a non trivial solution via minimization on the Nehari manifold.

8. Let Ω be an open and bounded subset of $\mathbb{R}^N$, with $N \geq 3$. Take $p \in (2, 2^*]$ and define

$$S_p = \inf_{\substack{u \in H^1_0(\Omega) \\ u \neq 0}} \frac{\int_\Omega |\nabla u|^2 \, dx}{(\int_\Omega |u|^p \, dx)^{2/p}}$$

This number is positive by the Sobolev embedding Theorem 1.2.1. Consider the functional $I : H^1_0(\Omega) \to \mathbb{R}$ defined by

$$I(u) = \frac{1}{2} \int_\Omega |\nabla u|^2 \, dx - \frac{1}{p} \int_\Omega |u|^p \, dx$$

and let $\mathcal{N}$ be the Nehari manifold associated to I. Prove that

$$\inf_{u \in \mathcal{N}} I(u) = \left(\frac{1}{2} - \frac{1}{p} \right) S_p^{\frac{p}{p-2}}.$$

9. Prove Theorem 2.6.6 by modifying the argument of Example 1.3.20.

10. Let $\Omega \subset \mathbb{R}^N$, with $N \geq 3$, be open and bounded, and let $1 < p < N$ and $p < r < q < p^*$. Consider the p-Laplacian problem

$$\begin{cases} -\Delta_p u = |u|^{r-2}u + |u|^{q-2}u & \text{in } \Omega, \\ u = 0 & \text{on } \partial\Omega. \end{cases}$$

Define the energy functional and the Nehari manifold for this problem, and prove the existence of a non trivial solution via minimization on the Nehari manifold.

2.8 Bibliographical Notes

- Section 2.1: The techniques presented in this section apply to much more general cases, for instance to quasilinear problems. The direct methods of the Calculus of

Variations, built upon the two notions of coercivity and weak lower semicontinuity have been developed enormously and allow one to deal with general functionals depending on x, u and ∇u. See Chap. I.1 in Struwe [45], or Giusti [21] for a vast description of these problems.

- Section 2.2: The abstract structure behind it is a particular case of the Ky Fan–Von Neumann Theorem. The content of the section is a simplified version of a result from Kavian [26]. Min–maximization under convexity and concavity assumptions is one of central themes of *convex analysis*. A comprehensive treatment of it, with applications to differential problems, can be found in the book [19] by Ekeland and Temam.

- Section 2.3: The failure of the direct method (minimization) for problems with superlinear growth due to the fact that the functionals are not bounded from below led to the idea of constraints. Some problems in mechanics and geometry were the main motivations for the introduction of artificial constraints. The Nehari manifold was introduced in Nehari [36] in the context of ordinary differential equations. A modern review on superlinear Dirichlet problems is Bartsch, Wang, and Willem [8].

- Section 2.4: The question whether "perturbations" can destroy existence results has been widely analyzed in the last few years, especially when the unperturbed problem possesses infinitely many solutions (as is the case for odd nonlinearities). The interested reader can read Rabinowitz [43], or Chap. II.7 of Struwe [45] and the references therein. The reading however requires more sophisticated techniques than those described in these notes.

- Section 2.5: The method of minimization on the Nehari manifold is particularly useful for nonhomogeneous nonlinearities, for which minimization on spheres does not apply. A rather detailed exposition can be found in the books by Willem [48], and Kuzin–Pohožaev [27]; the applications given in [27] however concern problems on unbounded domains, which we treat in Chap. 3.

- Section 2.6: There is by now a vast literature on problems involving the p-Laplacian. A good starting point is Garcia Azorero and Peral [20]. A systematic treatment of many questions concerning the p-Laplacian can be found in the book by Lindqvist [30].

Chapter 3
Minimization Techniques: Lack of Compactness

In this chapter we present some examples of problems where *compactness* is not guaranteed *a priori*. The lack of compactness can take different forms, but in the simplest case, it is manifest through the fact that minimizing sequences are maybe bounded, but not (pre-)compact in the function spaces where the problem is set.

The reasons for this often come from geometrical or physical aspects of the problem, and we refer for example to [45] for a discussion on this point of view.

Here we confine ourselves to some more or less simple examples of problems with lack of compactness and we try to show some ways to overcome the obstacle.

As in Chap. 2, we will study equations with different hypotheses on the nonlinearity f. In particular when dealing with critical problems (Sect. 3.4) we will study the homogeneous nonlinearity $f(t) = |t|^{2^*-2}t$ through minimization on spheres (as in Sect. 2.3.1), while in the first part of this section we will study a nonhomogeneous nonlinearity, applying the method of minimization on the Nehari manifold. In principle we could treat a generic f satisfying a set of hypotheses including $(\mathbf{f_1})$–$(\mathbf{f_4})$, but for the sake of simplicity we will deal with the specific and simple nonhomogeneous nonlinearity given by

$$f(t) = |t|^{p-2}t + \lambda|t|^{r-2}t$$

with $2 < r < p < 2^*$ and $\lambda \in \mathbb{R}$ (for $\lambda = 0$ this includes the homogeneous case).

3.1 A Radial Problem in $\mathbb{R}^N$

A first case of lack of compactness takes place when one works in *unbounded* open sets. For simplicity we treat the case where $\Omega = \mathbb{R}^N$, with $N \geq 3$, by studying the following problem:

$$\begin{cases} -\Delta u + \alpha u = |u|^{p-2}u + \lambda|u|^{r-2}u & \text{in } \mathbb{R}^N, \\ u \in H^1(\mathbb{R}^N) \end{cases} \tag{3.1}$$

where $\alpha > 0$, $\lambda \in \mathbb{R}$, $p, r \in (2, 2^*)$ and $p > r$.

M. Badiale, E. Serra, *Semilinear Elliptic Equations for Beginners*, Universitext, DOI 10.1007/978-0-85729-227-8_3, © Springer-Verlag London Limited 2011

A weak solution of (3.1) is a function $u \in H^1(\mathbb{R}^N)$ such that

$$\int_{\mathbb{R}^N} \nabla u \cdot \nabla v\, dx + \alpha \int_{\mathbb{R}^N} uv\, dx$$
$$= \int_{\mathbb{R}^N} |u|^{p-2} uv\, dx + \lambda \int_{\mathbb{R}^N} |u|^{r-2} uv\, dx \quad \forall v \in H^1(\mathbb{R}^N).$$

Remark 3.1.1 The condition $u \in H^1(\mathbb{R}^N)$ is a typical way to replace the boundary conditions on $\mathbb{R}^N$. Indeed, it can be viewed as a weak form of the requirement $u(x) \to 0$ for $|x| \to \infty$, a natural condition to associate to the equation in many cases. Notice however that $u \in H^1(\mathbb{R}^N)$ *does not* imply that u vanishes at infinity. Often the condition $u \in H^1(\mathbb{R}^N)$ *and* the fact that u solves the equation imply that $u(x) \to 0$ at infinity, but this has to be proved in each case.

A problem like (3.1) is not compact. Without supplying irrelevant details, imagine that it has a solution $u \not\equiv 0$ which minimizes some functional. Since the problem is invariant under translations, it is clear that for every $y \in \mathbb{R}^N$, the sequence $u_k(x) = u(x + ky)$ is a bounded (in $H^1(\mathbb{R}^N)$, for instance) minimizing sequence which cannot be precompact in any $L^p(\mathbb{R}^N)$. Thus, roughly speaking, the problem admits bounded minimizing sequences that do not converge. The reason for this is the invariance of $\mathbb{R}^N$ with respect to translations, which in turn makes the embedding of $H^1(\mathbb{R}^N)$ into $L^p(\mathbb{R}^N)$ not compact for any p.

A natural attempt to overcome this problem is to guess that translational invariance is the *only* reason for the failure of compactness, and to try to work in a space of functions where translations are not allowed. This is possible in this case because the problem is also invariant under rotations, so that one can try to work in spaces of *radial* functions. Let us make this precise.

We define

$$D_r = \{u \in D^{1,2}(\mathbb{R}^N) \mid u \text{ is radial}\}$$

and

$$H_r = \{u \in H^1(\mathbb{R}^N) \mid u \text{ is radial}\};$$

clearly $H_r \subset D_r$; see Sect. 1.2.1 for the definition of $D^{1,2}(\mathbb{R}^N)$. Also notice that by classical results (see [11]), functions in D_r and in H_r can be assumed to be continuous at all points except the origin.

We prove the following lemma, which gives a pointwise estimate for functions in D_r. The argument for its proof is a slight modification of a result in [37].

Lemma 3.1.2 *There exists a constant $C > 0$ such that for every $u \in D_r$,*

$$|u(x)| \leq C \frac{\|u\|}{|x|^{\frac{N-2}{2}}} \quad \text{for every } x \neq 0.$$

Proof Using a density argument, it is enough to prove the inequality for functions $u \in C_0^\infty(\mathbb{R}^N) \cap D_r$.

Let $\varphi : [0, +\infty[\to \mathbb{R}$ be defined by

$$\varphi(r) = u(x) \quad \text{if } r = |x|.$$

Then φ is C^∞ and has compact support in $[0, +\infty)$.

We now estimate, for $x \neq 0$,

$$|u(x)| = |\varphi(r)| = \left| \int_r^{+\infty} \varphi'(s)\,ds \right| \leq \int_r^{+\infty} |\varphi'(s)|\,ds = \int_r^{+\infty} |\varphi'(s)| s^{\frac{N-1}{2}} \frac{1}{s^{\frac{N-1}{2}}}\,ds$$

$$\leq \left(\int_r^{+\infty} |\varphi'(s)|^2 s^{N-1}\,ds \right)^{1/2} \left(\int_r^{+\infty} s^{1-N}\,ds \right)^{1/2}$$

$$= \frac{1}{\omega^{1/2}} \left(\omega \int_r^{+\infty} |\varphi'(s)|^2 s^{N-1}\,ds \right)^{1/2} \frac{1}{\sqrt{N-2}} \frac{1}{r^{\frac{N-2}{2}}}$$

$$= C \frac{\|u\|}{|x|^{\frac{N-2}{2}}},$$

where ω is the measure of the unit sphere in $\mathbb{R}^N$. $\qquad\square$

Remark 3.1.3 For functions $u \in H_r$ the following stronger decay estimate can be proved

$$|u(x)| \leq C \frac{\|u\|}{|x|^{\frac{N-1}{2}}} \quad \text{for every } |x| \geq 1,$$

see for example [2]. We won't need this stronger form.

We now prove that the heuristic idea that translations are the only obstruction to compactness is correct. The Sobolev inequalities on $\mathbb{R}^N$ (Theorem 1.2.2) show that H_r is continuously embedded into $L^p(\mathbb{R}^N)$, for $p \in [2, 2^*]$. The following lemma proves that this embedding is compact if $p \in (2, 2^*)$.

Lemma 3.1.4 *Let $p \in (2, 2^*)$. Then the embedding of H_r into $L^p(\mathbb{R}^N)$ is compact.*

Proof It is enough to show that if $\{u_k\}_k$ is a sequence in H_r such that $u_k \rightharpoonup 0$, then $u_k \to 0$ in $L^p(\mathbb{R}^N)$. Of course we can assume

$$u_k \to 0 \quad \text{in } L^p_{\text{loc}}(\mathbb{R}^N) \quad \text{and} \quad u_k(x) \to 0 \quad \text{a.e.}$$

Let $B = \{x \in \mathbb{R}^N \mid |x| < 1\}$ be the unit ball of $\mathbb{R}^N$, and let $B^c = \mathbb{R}^N \backslash B$. By what we have just said, as $k \to \infty$,

$$\int_{\mathbb{R}^N} |u_k|^p\,dx = \int_B |u_k|^p\,dx + \int_{B^c} |u_k|^p\,dx = \int_{B^c} |u_k|^p\,dx + o(1).$$

Let us fix $q > 2^*$. By Lemma 3.1.2, in B^c we have, for some appropriate constant $C > 0$,

$$|u_k(x)|^q \leq \frac{C}{|x|^{q\frac{N-2}{2}}},$$

and $q \frac{N-2}{2} > N$. This shows that

$$\frac{1}{|x|^{q\frac{N-2}{2}}} \in L^1(B^c),$$

and then, by dominated convergence,

$$\int_{B^c} |u_k(x)|^q \, dx \to 0.$$

Let now $t \in (0, 1)$ be such that $p = 2t + q(1 - t)$; by the Hölder inequality we can write

$$\int_{B^c} |u_k|^p \, dx = \int_{B^c} |u_k|^{2t} |u_k|^{q(1-t)} \, dx \le \left(\int_{B^c} |u_k|^2 \, dx \right)^t \left(\int_{B^c} |u_k|^q \, dx \right)^{1-t}.$$

Since the sequence $\{\int_{B^c} |u_k|^2 \, dx\}_k$ is bounded and $\int_{B^c} |u_k|^q \, dx \to 0$, we immediately obtain $\int_{B^c} |u_k|^p \, dx \to 0$, so that

$$\int_{\mathbb{R}^N} |u_k|^p \, dx \to 0.$$ □

Having established the main functional properties of the space H_r, we need a way to pass from functions in $H^1(\mathbb{R}^N)$ to functions in H_r. A way to do this, suitable for our purposes, is given by a procedure called Schwarz symmetrization. We can not give here an introduction to the theory of symmetrization, for which the reader can see [29]. We just describe roughly the idea for nonnegative functions and state the result that we will need in this section.

Let $u \in D^{1,2}(\mathbb{R}^N)$ be such that $u(x) \ge 0$ a.e. in $\mathbb{R}^N$. We denote, for $t > 0$,

$$\{u > t\} = \{x \in \mathbb{R}^N \mid u(x) > t\} \quad \text{and} \quad \mu(t) = |\{u > t\}|.$$

Notice that since $u \in L^{2^*}(\mathbb{R}^N)$, we have $\mu(t) < +\infty$ for all $t > 0$. The Schwarz symmetrization constructs a *radial* function $u^* : \mathbb{R}^N \to \mathbb{R}$ such that

$$\{u^* > t\} = B_{\rho(t)} \quad \text{with } |B_{\rho(t)}| = \mu(t).$$

Thus, the sets where u and u^* are greater than t have the same measure. It is easy to see that u^* is (radially) nonincreasing. The most important properties of u^* are stated in the following result.

Theorem 3.1.5 *For every $u \in D^{1,2}(\mathbb{R}^N)$, $u \ge 0$, there results $u^* \in D_r$, $u^* \ge 0$,*

$$\int_{\mathbb{R}^N} |\nabla u^*|^2 \, dx \le \int_{\mathbb{R}^N} |\nabla u|^2 \, dx \quad \text{and}$$

$$\int_{\mathbb{R}^N} |u^*|^p \, dx = \int_{\mathbb{R}^N} |u|^p \, dx, \quad \text{for all } p > 1.$$

These properties imply of course that if $u \in H^1(\mathbb{R}^N)$ then $u^* \in H_r$.

We can now state and prove the main result of this section.

Theorem 3.1.6 *Problem* (3.1) *admits at least one nontrivial nonnegative solution.*

This result will be obtained by a sequence of lemmas, and the scheme of the argument is the same as that we have used in the compact case.

We equip $H^1(\mathbb{R}^N)$ with the norm

$$\|u\|^2 = \int_{\mathbb{R}^N} |\nabla u|^2 \, dx + \alpha \int_{\mathbb{R}^N} |u|^2 \, dx$$

and we look for a solution as a minimizer of the associated functional constrained on the Nehari manifold. So we define the functional $I : H^1(\mathbb{R}^N) \to \mathbb{R}$ as

$$I(u) = \frac{1}{2} \int_{\mathbb{R}^N} |\nabla u|^2 \, dx + \frac{\alpha}{2} \int_{\mathbb{R}^N} u^2 \, dx - \frac{1}{p} \int_{\mathbb{R}^N} |u|^p \, dx - \frac{\lambda}{r} \int_{\mathbb{R}^N} |u|^r \, dx$$

$$= \frac{1}{2} \|u\|^2 - \frac{1}{p} |u|_p^p - \frac{\lambda}{r} |u|_r^r$$

and the Nehari Manifold

$$\mathcal{N} = \left\{ u \in H^1(\mathbb{R}^N) \mid u \neq 0, \, I'(u)u = 0 \right\}$$

$$= \left\{ u \in H^1(\mathbb{R}^N) \mid u \neq 0, \, \|u\|^2 = |u|_p^p + \lambda |u|_r^r \right\}.$$

Notice that, if $u \in \mathcal{N}$, then

$$I(u) = \left(\frac{1}{2} - \frac{1}{r} \right) \|u\|^2 + \left(\frac{1}{r} - \frac{1}{p} \right) |u|_p^p.$$

As $2 < r < p$, we have $I(u) > (\frac{1}{2} - \frac{1}{r}) \|u\|^2 > 0$ for all $u \in \mathcal{N}$.

Lemma 3.1.7 *The Nehari manifold is not empty.*

Proof We fix $u \in H^1(\mathbb{R}^N)$, with $u \neq 0$, and for $t \in \mathbb{R}$ we define the function

$$\gamma(t) = I'(tu)tu = t^2 \|u\|^2 - t^p |u|_p^p - \lambda t^r |u|_r^r.$$

It is then obvious that $\gamma(t) > 0$ for small $t > 0$ and that $\lim_{t \to +\infty} \gamma(t) = -\infty$; hence there exists $t > 0$ such that $I'(tu)tu = 0$, so that $tu \in \mathcal{N}$.   □

We define

$$m = \inf_{u \in \mathcal{N}} I(u)$$

and we try to show that m is attained by some $u \in \mathcal{N}$.

Lemma 3.1.8 *There results $m > 0$.*

Proof If $\lambda = 0$, the nonlinearity is homogeneous and the argument is the same as that of the compact case (Sect. 2.3.2). If $\lambda \neq 0$ and $u \in \mathcal{N}$, then

$$\|u\|^2 = |u|_p^p + \lambda |u|_r^r \leq C \left(\|u\|^p + \|u\|^r \right),$$

namely

$$1 \leq C\left(\|u\|^{p-2} + \|u\|^{r-2}\right).$$

If $\|u\| \leq 1$ this implies $1 \leq 2C\|u\|^{r-2}$.

So we have obtained, for all $u \in \mathcal{N}$,

$$\|u\| \geq \min\left\{1, (2C)^{-\frac{1}{r-2}}\right\}.$$

Therefore, if $u \in \mathcal{N}$ we have

$$I(u) > \left(\frac{1}{2} - \frac{1}{r}\right)\|u\|^2 \geq \left(\frac{1}{2} - \frac{1}{r}\right)\min\left\{1, (2C)^{-\frac{1}{r-2}}\right\}^2,$$

and the lemma is proved.   □

Lemma 3.1.9 *There exists $u \in \mathcal{N}$ such that $I(u) = m$.*

Proof We first show that we can take a minimizing sequence for m in $\mathcal{N} \cap H_r$. To this aim, let $\{v_k\}_k \subseteq \mathcal{N}$ be a minimizing sequence. As usual we can assume $v_k \geq 0$. Let $w_k = v_k^* \in H_r$ be the non negative radial function given by Theorem 3.1.5. We have

$$\|w_k\|^2 = \int_{\mathbb{R}^N} |\nabla v_k^*|^2\, dx + \alpha \int_{\mathbb{R}^N} |v_k^*|^2\, dx \leq \int_{\mathbb{R}^N} |\nabla v_k|^2\, dx + \alpha \int_{\mathbb{R}^N} |v_k|^2\, dx$$

$$= \int_{\mathbb{R}^N} |v_k|^p\, dx + \lambda \int_{\mathbb{R}^N} |v_k|^r\, dx = \int_{\mathbb{R}^N} |v_k^*|^p\, dx + \lambda \int_{\mathbb{R}^N} |v_k^*|^r\, dx$$

$$= \int_{\mathbb{R}^N} |w_k|^p\, dx + \lambda \int_{\mathbb{R}^N} |w_k|^r\, dx.$$

Hence if we set

$$\gamma(t) = I'(tw_k)tw_k = t^2\|w_k\|^2 - t^p|w_k|_p^p - \lambda|w_k|_r^r,$$

we have $\gamma(1) \leq 0$, while $\gamma(t) > 0$ for t positive and small. Therefore there exists $t_k \in (0, 1]$ such that $\gamma(t_k) = 0$, that is, $t_k w_k \in \mathcal{N}$. We obtain

$$m \leq I(t_k w_k)$$

$$= t_k^2\left(\frac{1}{2} - \frac{1}{r}\right)\|w_k\|^2 + t_k^p\left(\frac{1}{r} - \frac{1}{p}\right)|w_k|_p^p$$

$$\leq \left(\frac{1}{2} - \frac{1}{r}\right)\|w_k\|^2 + \left(\frac{1}{r} - \frac{1}{p}\right)|w_k|_p^p$$

$$\leq \left(\frac{1}{2} - \frac{1}{r}\right)\|v_k\|^2 + \left(\frac{1}{r} - \frac{1}{p}\right)|v_k|_p^p = I(v_k).$$

This implies that $\{t_k w_k\}_k$ is a minimizing sequence for m and $t_k w_k \in H_r$, as we had claimed. In the sequel we set $u_k = t_k w_k$. Of course, $u_k \geq 0$, and we can assume that, up to subsequences, $u_k \rightharpoonup u$ in $H^1(\mathbb{R}^N)$. By Lemma 3.1.4 we obtain

$$u_k \to u \quad \text{in } L^p(\mathbb{R}^N) \text{ and in } L^r(\mathbb{R}^N)$$

and, again up to subsequences, $u_k(x) \to u(x)$ almost everywhere, so that $u(x) \geq 0$ a.e. and $u \in H_r$. We now prove that the weak limit u belongs to $\mathcal{N}$ and $I(u) = m$. Let us first check that $u \in \mathcal{N}$. We have

$$0 < C \leq \|u_k\|^2 = |u_k|_p^p + \lambda |u_k|_r^r \tag{3.2}$$

and hence, passing to the limit,

$$0 < C \leq |u|_p^p + \lambda |u|_r^r;$$

this implies $u \neq 0$. Still from (3.2), we also get

$$\|u\|^2 \leq |u|_p^p + \lambda |u|_r^r.$$

If

$$\|u\|^2 = |u|_p^p + \lambda |u|_r^r,$$

then $u \in \mathcal{N}$. So, arguing by contradiction, we assume that

$$\|u\|^2 < |u|_p^p + \lambda |u|_r^r.$$

Defining as above, for $t > 0$,

$$\gamma(t) = I'(tu)tu = t^2 \|u\|^2 - t^p |u|_p^p - \lambda t^r |u|_r^r,$$

we see that $\gamma(t) > 0$ for small $t > 0$ while

$$\gamma(1) = \|u\|^2 - |u|_p^p - \lambda |u|_r^r < 0.$$

So there is $t \in (0, 1)$ such that $tu \in \mathcal{N}$. Hence,

$$
\begin{aligned}
0 < m \leq I(tu) &= \left(\frac{1}{2} - \frac{1}{r}\right)\|tu\|^2 + \left(\frac{1}{r} - \frac{1}{p}\right)|tu|_p^p \\
&= t^2 \left(\frac{1}{2} - \frac{1}{r}\right)\|u\|^2 + t^p \left(\frac{1}{r} - \frac{1}{p}\right)|u|_p^p \\
&< \left(\frac{1}{2} - \frac{1}{r}\right)\|u\|^2 + \left(\frac{1}{r} - \frac{1}{p}\right)|u|_p^p \\
&\leq \liminf_k \left(\frac{1}{2} - \frac{1}{r}\right)\|u_k\|^2 + \lim_k \left(\frac{1}{r} - \frac{1}{p}\right)|u_k|_p^p \\
&= \liminf_k \left(\left(\frac{1}{2} - \frac{1}{r}\right)\|u_k\|^2 + \left(\frac{1}{r} - \frac{1}{p}\right)|u_k|_p^p\right) = \liminf_k I(u_k) = m.
\end{aligned}
$$

This contradiction proves that $\|u\|^2 = |u|_p^p + \lambda |u|_r^r$, and therefore $u \in \mathcal{N}$. By the weak lower semicontinuity of the norm it is straightforward to deduce that $I(u) \leq \liminf_k I(u_k) = m$, and the lemma is proved. $\square$

Lemma 3.1.10 *The minimum u is a critical point for I in $H^1(\mathbb{R}^N)$.*

Proof Fix $v \in H^1(\mathbb{R}^N)$ and $\varepsilon > 0$ such that $u + sv \neq 0$ for all $s \in (-\varepsilon, \varepsilon)$. Define a function $\varphi : (-\varepsilon, \varepsilon) \times (0, +\infty) \to \mathbb{R}$ by

$$\varphi(s, t) = I'(t(u + sv))t(u + sv) = t^2 \|u + sv\|^2 - t^p |u + sv|_p^p - \lambda t^r |u + sv|_r^r.$$

Then $\varphi(0,1) = \|u^2\| - |u|_p^p - \lambda|u|_r^r = 0$ and

$$\frac{\partial\varphi}{\partial t}(0,1) = 2\|u\|^2 - p|u|_p^p - r\lambda|u|_r^r = (2-r)\|u\|^2 + (r-p)|u|_p^p < 0.$$

So, by the Implicit Function Theorem there exists a C^1 function $t : (-\varepsilon_0, \varepsilon_0) \to \mathbb{R}$ such that $t(0) = 1$ and

$$\varphi(s, t(s)) = 0$$

for all $s \in (-\varepsilon_0, \varepsilon_0)$. Defining

$$\gamma(s) = I(t(s)(u+sv)),$$

we see that the function γ is differentiable and has a minimum point at $s = 0$; therefore

$$0 = \gamma'(0) = I'(t(0)u)(t'(0)u + t(0)v) = I'(u)v.$$

Since this holds for all $v \in H^1(\mathbb{R}^N)$, we have $I'(u) = 0$. $\square$

Remark 3.1.11 In view of Lemma 3.1.2, the *radial* solution u found above does indeed satisfy $u(x) \to 0$ as $|x| \to \infty$.

3.2 A Problem with Unbounded Potential

In this section we study a variant of problem (3.1), where the constant α is replaced by a function $q(x)$ (the *potential* in the physical literature).

Precisely, we consider the problem

$$\begin{cases} -\Delta u + q(x)u = |u|^{p-2}u + \lambda|u|^{r-2}u & \text{in } \mathbb{R}^N, \\ u \in H^1(\mathbb{R}^N), \end{cases} \tag{3.3}$$

with $2 < r < p < 2^*$ and $\lambda \in \mathbb{R}$.

The fact that we are working on $\mathbb{R}^N$ suggests that we may have to face a lack of compactness. We will now describe what kind of assumptions on q allow us to overcome this type of problem.

We assume that $q : \mathbb{R}^N \to \mathbb{R}$ is continuous (for simplicity) and satisfies

($\mathbf{q_1}$) $\inf_{x \in \mathbb{R}^N} q(x) > 0$ and $\lim_{|x| \to +\infty} q(x) = +\infty$.

Notice that under ($\mathbf{q_1}$), if $u \in H^1(\mathbb{R}^N)$ a term like

$$\int_{\mathbb{R}^N} q(x)u^2\,dx$$

may diverge; this means that we cannot work directly in the space $H^1(\mathbb{R}^N)$.

Also, the function q is not necessarily radial, so that it makes no sense to restrict the problem to spaces of radial functions as we did in the previous section. Compactness will be recovered by means of assumption ($\mathbf{q_1}$).

We introduce the following subspace X of $H^1(\mathbb{R}^N)$:

$$X = \left\{ u \in H^1(\mathbb{R}^N) \,\middle|\, \int_{\mathbb{R}^N} q(x)u^2\,dx < +\infty \right\}.$$

It is easy to recognize that X is a Hilbert space with scalar product and norm given by

$$(u|v) = \int_{\mathbb{R}^N} \nabla u \cdot \nabla v\,dx + \int_{\mathbb{R}^N} q(x)uv\,dx \quad \text{and} \quad \|u\|^2 = (u|u).$$

A weak solution of Problem (3.3) is a function $u \in X$ such that

$$\int_{\mathbb{R}^N} \nabla u \cdot \nabla v\,dx + \int_{\mathbb{R}^N} q(x)uv\,dx$$
$$= \int_{\mathbb{R}^N} |u|^{p-2}uv\,dx + \lambda \int_{\mathbb{R}^N} |u|^{r-2}uv\,dx \tag{3.4}$$

for every $v \in X$.

From $(\mathbf{q_1})$ one deduces immediately that X is embedded continuously into $H^1(\mathbb{R}^N)$ and therefore also into $L^p(\mathbb{R}^N)$, for $p \in [2, 2^*]$. But much more can be said about this last embedding.

Lemma 3.2.1 *The embedding of X into $L^p(\mathbb{R}^N)$ is compact for all $p \in [2, 2^*)$.*

Proof Let $\{u_k\}_k \subset X$ be a sequence such that $u_k \rightharpoonup 0$ in X. We have to show that $u_k \to 0$ in $L^p(\mathbb{R}^N)$. Extracting subsequences if necessary, we can assume that

$$u_k \to 0 \quad \text{in } L^p_{\text{loc}}(\mathbb{R}^N) \quad \text{and} \quad u_k(x) \to 0 \quad \text{a.e. in } \mathbb{R}^N.$$

We first show that $u_k \to 0$ in $L^2(\mathbb{R}^N)$. For this, let

$$M = \sup_k \{\|u_k\|^2\}.$$

For every $\varepsilon > 0$, let $R_\varepsilon > 0$ be such that $q(x) \geq M/\varepsilon$ for all $|x| \geq R_\varepsilon$; this number exists due to $(\mathbf{q_1})$. We know that

$$\lim_{k\to\infty} \int_{|x|<R_\varepsilon} u_k^2\,dx = 0,$$

so that we obtain

$$\int_{\mathbb{R}^N} u_k^2\,dx = \int_{|x|<R_\varepsilon} u_k^2\,dx + \int_{|x|\geq R_\varepsilon} u_k^2\,dx = o(1) + \int_{|x|\geq R_\varepsilon} \frac{1}{q}qu_k^2\,dx$$
$$\leq o(1) + \frac{\varepsilon}{M}\int_{|x|\geq R_\varepsilon} qu_k^2\,dx \leq o(1) + \frac{\varepsilon}{M}\|u_k\|^2 \leq o(1) + \varepsilon.$$

This shows that

$$\limsup_{k\to\infty} \int_{\mathbb{R}^N} u_k^2\,dx \leq \varepsilon,$$

and since this holds for every $\varepsilon > 0$, we conclude that

$$\lim_{k \to \infty} \int_{\mathbb{R}^N} u_k^2 \, dx = 0.$$

We can now finish easily with the usual interpolation argument: the sequence u_k being bounded in X, is also bounded in $H^1(\mathbb{R}^N)$, and then also in $L^{2^*}(\mathbb{R}^N)$. Take any $p \in (2, 2^*)$ and write $p = 2t + 2^*(1 - t)$ with $t \in (0, 1)$; then

$$\int_{\mathbb{R}^N} |u_k|^p \, dx \leq \left(\int_{\mathbb{R}^N} |u_k|^2 \, dx \right)^t \left(\int_{\mathbb{R}^N} |u_k|^{2^*} \, dx \right)^{1-t} \to 0. \qquad \square$$

We can now introduce the Nehari manifold and repeat the argument of the previous section. We leave this to the reader, and we just state the existence result.

Theorem 3.2.2 *Let* $p, r \in (2, 2^*)$, $r < p$, *and assume that* $(\mathbf{q_1})$ *holds. Then there exists* $u \in X$, $u \geq 0$, *that solves weakly Problem* (3.3).

The solution that we have found satisfies (3.4). We want now to see that the weak equation is satisfied for any test function in $H^1(\mathbb{R}^N)$.

Theorem 3.2.3 *If* $u \in X$ *satisfies* (3.4), *then it also satisfies*

$$\int_{\mathbb{R}^N} \nabla u \cdot \nabla v \, dx + \int_{\mathbb{R}^N} q u v \, dx$$
$$= \int_{\mathbb{R}^N} |u|^{p-2} u v \, dx + \lambda \int_{\mathbb{R}^N} |u|^{r-2} u v \, dx \quad \forall v \in H^1(\mathbb{R}^N).$$

Proof Fix $v \in H^1(\mathbb{R}^N)$ and assume that $v \geq 0$. Let $\{\varphi_k\}_k$ be a sequence of C^∞ functions such that

- $0 \leq \varphi_k(x) \leq 1$ for every x,
- $\varphi_k(x) = 1$ if $|x| \leq k$,
- $\varphi_k(x) = 0$ if $|x| \geq k + 1$,
- $|\nabla \varphi_k(x)| \leq C$ for every x.

It is then easy to check that $\varphi_k(x)v \in H^1(\mathbb{R}^N)$ and that

$$\varphi_k(x)v \to v \quad \text{in } H^1(\mathbb{R}^N),$$

and then also in $L^p(\mathbb{R}^N)$ and $L^r(\mathbb{R}^N)$. As q is locally bounded and $\varphi_k v$ has compact support, we obtain that $\varphi_k v \in X$; therefore

$$\int_{\mathbb{R}^N} \nabla u \cdot \nabla(\varphi_k v) \, dx + \int_{\mathbb{R}^N} q u \varphi_k v \, dx$$
$$= \int_{\mathbb{R}^N} |u|^{p-2} u \varphi_k v \, dx + \lambda \int_{\mathbb{R}^N} |u|^{r-2} u \varphi_k v \, dx. \qquad (3.5)$$

Strong convergence in $H^1(\mathbb{R}^N)$ implies

$$\lim_k \int_{\mathbb{R}^N} \nabla u \cdot \nabla(\varphi_k v)\,dx = \int_{\mathbb{R}^N} \nabla u \cdot \nabla v\,dx,$$

$$\lim_k \int_{\mathbb{R}^N} |u|^{p-2} u \varphi_k v\,dx = \int_{\mathbb{R}^N} |u|^{p-2} uv\,dx,$$

$$\lim_k \int_{\mathbb{R}^N} |u|^{r-2} u \varphi_k v\,dx = \int_{\mathbb{R}^N} |u|^{r-2} uv\,dx,$$

and hence from (3.5) we deduce that

$$0 \le \lim_k \int_{\mathbb{R}^N} q u \varphi_k v\,dx$$

$$= -\int_{\mathbb{R}^N} \nabla u \cdot \nabla v\,dx + \int_{\mathbb{R}^N} |u|^{p-2} uv\,dx + \lambda \int_{\mathbb{R}^N} |u|^{r-2} uv\,dx < +\infty.$$

On the other hand we have

$$0 \le q(x)\varphi_k(x)v(x) \le q(x)\varphi_{k+1}(x)v(x)$$

for all k and almost every x so that by Beppo Levi's monotone convergence theorem, we have

$$\lim_k \int_{\mathbb{R}^N} q u \varphi_k v\,dx = \int_{\mathbb{R}^N} q u v\,dx.$$

Summing up, we obtain

$$\int_{\mathbb{R}^N} q u v\,dx = -\int_{\mathbb{R}^N} \nabla u \cdot \nabla v\,dx + \int_{\mathbb{R}^N} |u|^{p-2} uv\,dx + \lambda \int_{\mathbb{R}^N} |u|^{r-2} uv\,dx,$$

and this holds for $v \in H^1(\mathbb{R}^N)$ and $v \ge 0$. To get the result for any v it is enough to write $v = v^+ - v^-$, and to do the computations for v^+ and v^- separately. $\qquad\square$

3.3 A Serious Loss of Compactness

In this section we study a problem where the loss of compactness takes a more serious form. The method to overcome the problem will therefore be considerably longer and more involved than in the preceding sections.

We study the same equation as in Sect. 3.2, namely

$$\begin{cases} -\Delta u + q(x)u = |u|^{p-2}u + \lambda|u|^{r-2}u & \text{in } \mathbb{R}^N, \\ u \in H^1(\mathbb{R}^N) \end{cases} \tag{3.6}$$

with $p, r \in (2, 2^*)$, $r < p$ and $\lambda \in \mathbb{R}$ but this time we make the following assumptions on q:

($\mathbf{q_2}$) q is continuous and $0 < \delta := \inf_{\mathbb{R}^N} q < \sup_{\mathbb{R}^N} q =: \alpha < +\infty$.
($\mathbf{q_3}$) $\lim_{|x| \to +\infty} q(x) = \alpha$.

We will work in $H^1(\mathbb{R}^N)$ with scalar product and norm given by

$$(u|v) = \int_{\mathbb{R}^N} \nabla u \cdot \nabla v\,dx + \int_{\mathbb{R}^N} q(x)uv\,dx \quad \text{and} \quad \|u\|^2 = (u|u).$$

Thanks to $(\mathbf{q_2})$, these are equivalent to the standard scalar product and norm of $H^1(\mathbb{R}^N)$.

A weak solutions of problem (3.6) is a function $u \in H^1(\mathbb{R}^N)$ such that

$$\int_{\mathbb{R}^N} \nabla u \cdot \nabla v \, dx + \int_{\mathbb{R}^N} quv \, dx$$
$$= \int_{\mathbb{R}^N} |u|^{p-2} uv \, dx + \lambda \int_{\mathbb{R}^N} |u|^{r-2} uv \, dx \quad \forall v \in H^1(\mathbb{R}^N),$$

namely a critical point of the functional $I \in C^1(H^1(\mathbb{R}^N))$ defined by

$$I(u) = \frac{1}{2} \int_{\mathbb{R}^N} \left(|\nabla u|^2 \, dx + q(x)u^2 \right) dx - \frac{1}{p} \int_{\mathbb{R}^N} |u|^p \, dx - \frac{\lambda}{r} \int_{\mathbb{R}^N} |u|^r \, dx$$
$$= \frac{1}{2} \|u\|^2 - \frac{1}{p} |u|_p^p - \frac{\lambda}{r} |u|_r^r \, dx.$$

Remark 3.3.1 The loss of compactness appears in the same way as in the preceding example, where it was overcome thanks to the behavior of q at infinity (assumption $(\mathbf{q_1})$).

In the present case, the behavior of q specified by assumption $(\mathbf{q_1})$ prevents the use of the same argument, and this is what makes the problem considerably more difficult.

We are going to prove the following result through a series of lemmas, in which we tacitly assume that $(\mathbf{q_2})$ and $(\mathbf{q_3})$ hold.

Theorem 3.3.2 *Let* $p, r \in (2, 2^*)$, $r < p$, *and assume that* $(\mathbf{q_2})$ *and* $(\mathbf{q_3})$ *hold. Then Problem (3.6) admits at least one nontrivial and nonnegative (weak) solution.*

As in previous sections, we use the method of minimization on the Nehari manifold. We set again

$$\mathcal{N} = \left\{ u \in H^1(\mathbb{R}^N) \mid u \neq 0, \, I'(u)u = 0 \right\}$$
$$= \left\{ u \in H^1(\mathbb{R}^N) \mid u \neq 0, \, \|u\|^2 = |u|_p^p + \lambda |u|_r^r \right\}.$$

If $u \in \mathcal{N}$,

$$I(u) = \left(\frac{1}{2} - \frac{1}{r} \right) \|u\|^2 + \left(\frac{1}{r} - \frac{1}{p} \right) |u|_p^p,$$

and since $2 < r < p$, then

$$I(u) > \left(\frac{1}{2} - \frac{1}{r} \right) \|u\|^2 > 0$$

for all $u \in \mathcal{N}$.

We define

$$m = \inf_{u \in \mathcal{N}} I(u).$$

Using the same arguments as above one can immediately prove the following result.

Lemma 3.3.3 *The Nehari manifold is not empty,* $\inf_{u \in \mathcal{N}} \|u\| > 0$ *and* $m > 0$.

We consider now a minimizing sequence $\{u_k\}_k \subset \mathcal{N}$. Of course we can assume $u_k(x) \geq 0$ almost everywhere in $\mathbb{R}^N$. The sequence u_k is bounded in $H^1(\mathbb{R}^N)$, and therefore, up to subsequences, there exists $u \in H^1(\mathbb{R}^N)$ such that

- $u_k \rightharpoonup u$ in $H^1(\mathbb{R}^N), L^p(\mathbb{R}^N), L^r(\mathbb{R}^N)$,
- $u_k \to u$ in $L^p_{\text{loc}}(\mathbb{R}^N)$,
- $u_k(x) \to u(x)$ a.e.

The last property tells us at once that $u \geq 0$. By extracting a further subsequence, if necessary, we can define $\beta \geq 0$ as

$$\beta = \lim_k \int_{\mathbb{R}^N} |u_k|^p$$

and we let

$$l = \int_{\mathbb{R}^N} |u|^p \, dx;$$

by weak convergence it is obvious that $l \in [0, \beta]$.

Lemma 3.3.4 *There results* $\beta > 0$.

Proof If $\beta = 0$, we have

$$\lim_k |u_k|_p = 0.$$

Let $t \in (0, 1)$ be such that $r = 2t + (1 - t)p$. Then by the Hölder inequality,

$$\int_{\mathbb{R}^N} |u_k|^r \, dx = \int_{\mathbb{R}^N} |u_k|^{2t} |u_k|^{p(1-t)} \, dx \leq |u_k|_2^{2t} |u_k|_p^{p(1-t)}.$$

As $\{|u_k|_2\}_k$ is a bounded sequence, we see that

$$\lim_k |u_k|_r = 0.$$

Since $u_k \in \mathcal{N}$, $\|u_k\|^2 = |u_k|_p^p + \lambda |u_k|_r^r$, and then we obtain $u_k \to 0$ in $H^1(\mathbb{R}^N)$ and $m = 0$, a contradiction. $\qquad \square$

Remark 3.3.5 Up to now the proof has followed the same lines that we have encountered many times. The real difficulty, the point where the lack of compactness becomes manifest, is to show that the minimizing sequence converges in a strong enough sense to pass to the limit in the nonlinear terms. In the preceding sections, compactness was restored from the beginning, either by working in the space of radial functions, or by the use of the space X, as testified by Lemmas 3.1.4 and 3.2.1. In the present case, nothing of the sort can be used, and this is what makes the present problem more difficult. The main argument of the proof consists in a rather *careful analysis* of the minimizing sequences, much more precise than the ones seen so far. We will try to classify the possible behaviors of minimizing sequences, and from this we will first rule out some possibilities, and then show that in the remaining cases there is enough compactness to conclude the proof.

This kind of argument is a first (and simplified) example of the application of a highly refined theory aimed at classifying *all* the possible behaviors of sequences of "approximate solutions", like minimizing sequences. The theory, beyond the scopes of this book, was developed by P.-L. Lions in a series of papers [31–34], and culminates with the "Concentration–Compactness Principle", a largely used tool in problems with lack of compactness.

We will find another example of application of this principle in the next section.

For the analysis of the minimizing sequence $\{u_k\}$ that we need here it is sufficient to consider three cases: $l = \beta$, $l = 0$, and $l \in (0, \beta)$.

It is quite useful, in this and in many other cases where compactness is the problem, to visualize intuitively the minimizing sequence $\{u_k\}$ as made of number of "bumps" in $\mathbb{R}^N$, changing their shape and moving around as $k \to \infty$. One says that the "mass" of u_k, for example some integral of $|u_k|^p$, is concentrated where the bumps are located. Following this image, one can easily realize that if at least one bump escapes to infinity, then u_k will not be compact (in the weak limit some mass is lost, and there is no strong convergence in L^p for instance). This is of course not the only possible reason for a failure of compactness, but it certainly helps in following the technical parts of the argument.

For example, keeping in mind the definition of β and l, it is reasonable to visualize the case $l = \beta$ as a situation where *no* mass is lost, and hence compactness should be expected. The case $l = 0$ corresponds to a *total* loss of mass (all the bumps of u_k escape to infinity); if this happens, the key will be the analysis of an auxiliary problem, where $q(x)$ is replaced by $\alpha = \lim_{|x| \to \infty} q(x)$. The hardest case is $l \in (0, \beta)$ which means that only *some* of the mass is lost in the limit.

We now make all this rigorous, beginning from the simplest case and confirming our initial guess that compactness can be recovered.

Lemma 3.3.6 *If $l = \beta$, then $u \in \mathcal{N}$ and $I(u) = m$.*

Proof If $l = \beta$, in addition to $u_k \rightharpoonup u$ in $L^p(\mathbb{R}^N)$ we also have $|u_k|_p \to |u|_p$, and this implies $u_k \to u$ in $L^p(\mathbb{R}^N)$ (see [11]). By the same interpolation argument of the previous lemma, this yields also $u_k \to u$ in $L^r(\mathbb{R}^N)$. Hence, by weak convergence, we obtain

$$I(u) \leq \liminf_k I(u_k) = m,$$

while the relation $\|u_k\|^2 = |u_k|_p^p + \lambda |u_k|_r^r$ implies

$$\|u\|^2 \leq |u|_p^p + \lambda |u|_r^r.$$

If equality holds, then $u \in \mathcal{N}$ (recall that $l = \beta > 0$, so $u \neq 0$) and the lemma is proved.

The case $\|u\|^2 < |u|_p^p + \lambda |u|_r^r$ cannot occur, since arguing as usual, and taking $t \in (0, 1)$ such that $tu \in \mathcal{N}$, we would have

$$m \leq I(tu)$$
$$= t^2 \left(\frac{1}{2} - \frac{1}{r}\right) \|u\|^2 + t^p \left(\frac{1}{r} - \frac{1}{p}\right) |u|_p^p$$
$$< \left(\frac{1}{2} - \frac{1}{r}\right) \|u\|^2 + \left(\frac{1}{r} - \frac{1}{p}\right) |u|_p^p \leq \liminf_k I(u_k) = m,$$

a contradiction. $\qquad\qquad\qquad\qquad\qquad\qquad\qquad\qquad\qquad\qquad\square$

We now turn to the case $l = 0$. We are going to use some of the results of Sect. 3.1. First we introduce in $H^1(\mathbb{R}^N)$ the norm

$$\|u\|_\alpha^2 = \int_{\mathbb{R}^N} |\nabla u|^2 \, dx + \alpha \int_{\mathbb{R}^N} |u|^2 \, dx,$$

where α is the positive number defined in ($\mathbf{q_2}$). Of course $\|u\|_\alpha$ in an Hilbertian norm equivalent to $\|u\|$. Then we define a functional

$$I_\alpha : H^1(\mathbb{R}^N) \to \mathbb{R} \quad \text{as} \quad I_\alpha(u) = \frac{1}{2} \|u\|_\alpha^2 - \frac{1}{p} |u|_p^p - \lambda \frac{1}{r} |u|_r^r,$$

and the associated Nehari manifold

$$\mathcal{N}_\alpha = \left\{u \in H^1(\mathbb{R}^N) \mid u \neq 0, I_\alpha'(u)u = 0\right\}$$
$$= \left\{u \in H^1(\mathbb{R}^N) \mid u \neq 0, \|u\|_\alpha^2 = |u|_p^p + \lambda |u|_r^r\right\}.$$

Finally we define

$$m_\alpha = \inf_{u \in \mathcal{N}_\alpha} I_\alpha(u).$$

The intuitive idea behind the introduction of this auxiliary problem is that if $l = 0$, then all the mass of u_k escapes to infinity. But close to infinity $q(x)$ "looks like α", which is $\sup_{\mathbb{R}^N} q$: it should not be convenient, if u_k has to minimize I, to have its mass located where q is large. Let us see that this is indeed true. First we compare m and m_α.

Lemma 3.3.7 *There results $m < m_\alpha$.*

Proof By the results of Sect. 3.1 we know that $\mathcal{N}_\alpha \neq \emptyset$, $m_\alpha > 0$ and that there exists $u_0 \in \mathcal{N}_\alpha$ such that $I_\alpha(u_0) = m_\alpha$. We also know that $u_0 \geq 0$.

By ($\mathbf{q_2}$) and ($\mathbf{q_3}$) we infer that there exist $\delta_1 > 0$ and a ball $B_R(x_1)$ such that

$$q(x) \leq \alpha - \delta_1 \quad \forall x \in B_R(x_1).$$

Since u_0 does not vanish identically, there exist $\delta_2 > 0$, a ball $B_R(x_2)$ and a set $A \subseteq B_R(x_2)$ of positive measure such that

$$u_0(x) \geq \delta_2 \quad \text{a.e. in } A.$$

We now define a function $u_1 \in H^1(\mathbb{R}^N)$ as

$$u_1(x) = u_0(x - x_1 + x_2).$$

By the invariance of integrals with respect to translations we have

$$I_\alpha(u_1) = I_\alpha(u_0) = m_\alpha \quad \text{and} \quad I'_\alpha(u_1)u_1 = I'_\alpha(u_0)u_0 = 0.$$

Notice that if $x \in B_R(x_1)$, then $x - x_1 + x_2 \in B_R(x_2)$; therefore $u_1(x) \geq \delta_2$ a.e. in a set $A' \subseteq B_R(x_1)$ of positive measure. Then

$$\int_{B_R(x_1)} (\alpha - q(x))u_1^2 \, dx \geq \int_{A'} \delta_1 \delta_2^2 = C\delta_1 \delta_2^2,$$

where C is the measure of A'.

Since $q(x) \leq \alpha$ for every x, we obtain the estimate

$$\int_{\mathbb{R}^N} (\alpha - q(x))u_1^2 \, dx \geq \int_{B_R(x_1)} (\alpha - q(x))u_1^2 \, dx \geq C\delta_1 \delta_2^2 > 0,$$

so that

$$\int_{\mathbb{R}^N} q(x)u_1^2 \, dx < \alpha \int_{\mathbb{R}^N} u_1^2 \, dx,$$

which implies

$$\int_{\mathbb{R}^N} |\nabla u_1|^2 \, dx + \int_{\mathbb{R}^N} q(x)u_1^2 \, dx < \int_{\mathbb{R}^N} |\nabla u_1|^2 \, dx + \alpha \int_{\mathbb{R}^N} u_1^2 \, dx,$$

that is,

$$\|u_1\|^2 < \|u_1\|_\alpha^2.$$

We then have

$$I'(u_1)u_1 = \|u_1\|^2 - |u_1|_p^p - \lambda |u_1|_r^r < \|u_1\|_\alpha^2 - |u_1|_p^p - \lambda |u_1|_r^r$$
$$= I'_\alpha(u_1)u_1 = I'_\alpha(u_0)u_0 = 0.$$

Hence, by usual arguments, there exists $t \in (0, 1)$ such that $tu_1 \in \mathcal{N}$, so that

$$m \leq I(tu_1) = t^2 \left(\frac{1}{2} - \frac{1}{r} \right) \|u_1\|^2 + t^p \left(\frac{1}{r} - \frac{1}{p} \right) |u_1|_p^p$$

$$< \left(\frac{1}{2} - \frac{1}{r} \right) \|u_1\|^2 + \left(\frac{1}{r} - \frac{1}{p} \right) |u_1|_p^p$$

$$< \left(\frac{1}{2} - \frac{1}{r} \right) \|u_1\|_\alpha^2 + \left(\frac{1}{r} - \frac{1}{p} \right) |u_1|_p^p$$

$$= \left(\frac{1}{2} - \frac{1}{r} \right) \|u_0\|_\alpha^2 + \left(\frac{1}{r} - \frac{1}{p} \right) |u_0|_p^p = I_\alpha(u_0) = m_\alpha. \qquad \square$$

Now the last step: if *all* the mass of u_k escapes to infinity, then $I(u_k)$ is too large (at least m_α), and u_k cannot be a minimizing sequence.

Lemma 3.3.8 *The case $l = 0$ cannot occur.*

Proof If $l = 0$, then $u = 0$, which implies in particular that $u_k \to 0$ in $L^2_{\text{loc}}(\mathbb{R}^N)$. We first prove the following claim:

$$\lim_{k \to +\infty} \int_{\mathbb{R}^N} |q(x) - \alpha| u_k^2 \, dx = 0. \tag{3.7}$$

To prove (3.7) we fix $\varepsilon > 0$ and we take $R_\varepsilon > 0$ such that

$$|q(x) - \alpha| \leq \varepsilon \quad \forall |x| \geq R_\varepsilon;$$

this is possible by $(\mathbf{q_3})$. We can then estimate

$$\int_{\mathbb{R}^N} |q(x) - \alpha| u_k^2 \, dx = \int_{|x| \leq R_\varepsilon} |q(x) - \alpha| u_k^2 \, dx + \int_{|x| > R_\varepsilon} |q(x) - \alpha| u_k^2 \, dx$$

$$\leq C \int_{|x| \leq R_\varepsilon} u_k^2 \, dx + M\varepsilon,$$

where

$$C = \max_{x \in \mathbb{R}^N} |q(x) - \alpha| \quad \text{and} \quad M = \sup_k \int_{\mathbb{R}^N} u_k^2 \, dx.$$

When $k \to \infty$ we obtain

$$\limsup_k \int_{\mathbb{R}^N} |q(x) - \alpha| u_k^2 \, dx \leq \varepsilon M$$

for every $\varepsilon > 0$, because $u_k \to 0$ in $L^2_{\text{loc}}(\mathbb{R}^N)$. As this holds for every $\varepsilon > 0$, (3.7) is proved.

From (3.7) we deduce that

$$\lim_{k \to +\infty} \|u_k\|_\alpha = \lim_{k \to +\infty} \|u_k\|. \tag{3.8}$$

Now we know that

$$\|u_k\|_\alpha^2 \geq \|u_k\|^2 = |u_k|_p^p + \lambda |u_k|_r^r,$$

so that, by usual arguments, we can prove that for each k there is $t_k \geq 1$ such that $t_k u_k \in \mathcal{N}_\alpha$, namely

$$t_k^2 \|u_k\|_\alpha^2 = t_k^p |u_k|_p^p + \lambda t_k^r |u_k|_r^r. \tag{3.9}$$

We know that the sequences $\{|u_k|_p\}$, $\{|u_k|_r\}$ and $\{\|u_k\|_\alpha\}$ are bounded and that

$$\lim_k |u_k|_p^p = \beta > 0.$$

Therefore from (3.9) we deduce that $\{t_k\}_k$ is bounded and, up to a subsequence, we can assume $t_k \to t_0$ with, of course, $t_0 \geq 1$. We also know that

$$\|u_k\|_\alpha^2 = \|u_k\|^2 + \varepsilon_k,$$

with $\varepsilon_k \to 0$, and that

$$\lambda |u_k|_r^r = \|u_k\|^2 - |u_k|_p^p.$$

Substituting in (3.9) we get

$$\left(t_k^2 - t_k^r\right) \|u_k\|^2 + t_k \varepsilon_k = \left(t_k^p - t_k^r\right) |u_k|_p^p.$$

Passing to the limit, and setting $\gamma = \lim_k \|u_k\|^2 > 0$, we obtain

$$\left(t_0^2 - t_0^r\right) \gamma = \left(t_0^p - t_0^r\right) \beta;$$

so $t_0 > 1$ gives a contradiction because $2 < r < p$, and then necessarily $t_0 = 1$. Now we can write

$$m_\alpha \le I_\alpha(t_k u_k) = t_k^2 \left(\frac{1}{2} - \frac{1}{r}\right) \|u_k\|_\alpha^2 + t_k^p \left(\frac{1}{r} - \frac{1}{p}\right) |u_k|_p^p$$

$$= t_k^2 \left(\frac{1}{2} - \frac{1}{r}\right) \|u_k\|^2 + t_k^2 \left(\frac{1}{2} - \frac{1}{r}\right) \varepsilon_k + t_k^2 \left(\frac{1}{r} - \frac{1}{p}\right) |u_k|_p^p$$

$$+ \left(t_k^p - t_k^2\right) \left(\frac{1}{r} - \frac{1}{p}\right) |u_k|_p^p$$

$$= t_k^2 I(u_k) + o(1).$$

Passing to the limit we then obtain

$$m_\alpha \le m,$$

a contradiction which concludes the proof.　　　　　　　　　　　　　　　□

As we anticipated, the hard case occurs when $l \in (0, \beta)$, because this means that the weak limit u is not identically zero, namely that only part of the mass escapes to infinity. A careful analysis will show that this is enough to conclude. We define

$$l_1 = |u|_r^r.$$

We begin by choosing an increasing sequence $\{R_j\}_j$ of positive numbers such that $R_{j+1} > R_j + 1$, so that in particular $R_j \to +\infty$, and such that the following properties hold:

- $l - \frac{1}{j} \le \int_{B_{R_j}} |u|^p \, dx \le l$, hence also $\int_{B_{R_j}^c} |u|^p \, dx \le \frac{1}{j}$,
- $l_1 - \frac{1}{j} \le \int_{B_{R_j}} |u|^r \, dx \le l_1$, hence also $\int_{B_{R_j}^c} |u|^r \, dx \le \frac{1}{j}$,
- $\int_{B_{R_j}^c} u^2 \, dx \le \frac{1}{j}$.

Then we define

$$C_j = B_{R_j+1} \backslash B_{R_j} = \{x \in \mathbb{R}^N \mid R_j \le |x| < R_j + 1\}.$$

Since we have compactness on bounded subsets of $\mathbb{R}^N$, for every j there results

$$\lim_k \int_{B_{R_j}} |u_k|^p \, dx = \int_{B_{R_j}} |u|^p \, dx, \qquad \lim_k \int_{C_j} |u_k|^p \, dx = \int_{C_j} |u|^p \, dx,$$

$$\lim_k \int_{B_{R_j}} |u_k|^r \, dx = \int_{B_{R_j}} |u|^r \, dx, \qquad \lim_k \int_{C_j} |u_k|^r \, dx = \int_{C_j} |u|^r \, dx$$

and

$$\lim_k \int_{C_j} |u_k|^2 \, dx = \int_{C_j} |u|^2 \, dx.$$

Moreover, since

$$\int_{C_j} |u|^p \, dx \le \int_{B_{R_j}^c} |u|^p \, dx \le \frac{1}{j}, \qquad \int_{C_j} |u|^r \, dx \le \int_{B_{R_j}^c} |u|^r \, dx \le \frac{1}{j}$$

and

$$\int_{C_j} |u|^2 \, dx \le \int_{B_{R_j}^c} |u|^2 \, dx \le \frac{1}{j},$$

we can extract a subsequence u_{k_j} in order to have, for every $j \in \mathbb{N}$,

$$l - \frac{2}{j} \le \int_{B_{R_j}} |u_{k_j}|^p \, dx \le l + \frac{1}{j}, \qquad \int_{C_j} |u_{k_j}|^p \, dx \le \frac{2}{j},$$

$$l_1 - \frac{2}{j} \le \int_{B_{R_j}} |u_{k_j}|^r \, dx \le l_1 + \frac{1}{j}, \qquad \int_{C_j} |u_{k_j}|^r \, dx \le \frac{2}{j}, \qquad \int_{C_j} |u_{k_j}|^2 \, dx \le \frac{2}{j}.$$

We take $\{u_{k_j}\}$ as a new minimizing sequence renaming it $\{u_j\}_j$.

We also consider, for every j, a function $\psi_j \in C^\infty(\mathbb{R}^N)$ such that

- $0 \le \psi_j(x) \le 1$ for every x,
- $\psi_j(x) = 1$ if $|x| \le R_j$,
- $\psi_j(x) = 0$ if $|x| \ge R_j + 1$,
- $|\nabla \psi_j(x)| \le C$ for every x

and we define auxiliary functions

$$u'_j = \psi_j u_j, \quad \text{and} \quad u''_j = (1 - \psi_j) u_j.$$

Of course, $u'_j, u''_j \ge 0$ and $u_j = u'_j + u''_j$ for every j.

Lemma 3.3.9 *The following properties hold as $j \to \infty$:*

1. *u'_j tends to u weakly in $H^1(\mathbb{R}^N)$ and strongly in $L^p(\mathbb{R}^N)$ and in $L^r(\mathbb{R}^N)$, while u''_j tends to 0 weakly in $H^1(\mathbb{R}^N)$.*
2. *$\int_{\mathbb{R}^N} |u_j|^p \, dx = \int_{\mathbb{R}^N} |u'_j|^p \, dx + \int_{\mathbb{R}^N} |u''_j|^p \, dx + o(1)$.*
3. *$\int_{\mathbb{R}^N} |u_j|^r \, dx = \int_{\mathbb{R}^N} |u'_j|^r \, dx + \int_{\mathbb{R}^N} |u''_j|^r \, dx + o(1)$.*
4. *$\|u_j\|^2 \ge \|u'_j\|^2 + \|u''_j\|^2 + o(1)$.*

Proof 1. If B is any bounded set in $\mathbb{R}^N$, we know that $u_j = u'_j$ in B, for every j big enough, from which we can immediately conclude that $u'_j \rightharpoonup u$ in $H^1(\mathbb{R}^N)$ (hence in $L^p(\mathbb{R}^N)$ and $L^r(\mathbb{R}^N)$). This also implies $u''_j \rightharpoonup 0$ in $H^1(\mathbb{R}^N)$, $L^p(\mathbb{R}^N)$ and $L^r(\mathbb{R}^N)$.

Moreover,

$$\int_{\mathbb{R}^N} |u'_j|^p \, dx \geq \int_{B_{R_j}} |u'_j|^p \, dx = \int_{B_{R_j}} |u_j|^p \, dx \geq l - \frac{2}{j},$$

and on the other hand, recalling that $0 \leq u'_j \leq u_j$,

$$\int_{\mathbb{R}^N} |u'_j|^p \, dx = \int_{B_{R_j+1}} |u'_j|^p \, dx \leq \int_{B_{R_j+1}} |u_j|^p \, dx$$

$$= \int_{B_{R_j}} |u_j|^p \, dx + \int_{C_j} |u_j|^p \, dx \leq l + \frac{1}{j} + \frac{2}{j} = l + \frac{3}{j}.$$

Thus we obtain that

$$\lim_j \int_{\mathbb{R}^N} |u'_j|^p \, dx = l = \int_{\mathbb{R}^N} |u|^p \, dx,$$

and since we have weak convergence we can conclude (see e.g. [11]) that $u'_j \to u$ in $L^p(\mathbb{R}^N)$. In the same way we get $u'_j \to u$ in $L^r(\mathbb{R}^N)$.

2. It is immediate to check that $|u_j|^p \geq |u'_j|^p + |u''_j|^p$; hence,

$$\int_{\mathbb{R}^N} \left(|u_j|^p - |u'_j|^p - |u''_j|^p \right) dx \geq 0.$$

On the other hand,

$$\int_{\mathbb{R}^N} |u_j|^p \, dx = \int_{B_{R_j}} |u_j|^p \, dx + \int_{C_j} |u_j|^p \, dx + \int_{B^c_{R_j+1}} |u_j|^p \, dx$$

$$= \int_{B_{R_j}} |u'_j|^p \, dx + \int_{C_j} |u_j|^p \, dx + \int_{B^c_{R_j+1}} |u''_j|^p \, dx$$

$$\leq \int_{\mathbb{R}^N} |u'_j|^p \, dx + \int_{C_j} |u_j|^p \, dx + \int_{\mathbb{R}^N} |u''_j|^p \, dx$$

$$\leq \int_{\mathbb{R}^N} |u'_j|^p \, dx + \frac{2}{j} + \int_{\mathbb{R}^N} |u''_j|^p \, dx,$$

so that

$$0 \leq \int_{\mathbb{R}^N} \left(|u_j|^p - |u'_j|^p - |u''_j|^p \right) dx \leq \frac{2}{j},$$

which proves 2.

3. The identity for the L^r norms can be proved as in 2.

4. Exactly as above, $|u_j|^2 \geq |u'_j|^2 + |u''_j|^2$, and since $q(x) \geq 0$, we obtain

$$\int_{\mathbb{R}^N} q(x) u_j^2 \, dx \geq \int_{\mathbb{R}^N} q(x) |u'_j|^2 \, dx + \int_{\mathbb{R}^N} q(x) |u''_j|^2 \, dx.$$

Also,

$$\int_{\mathbb{R}^N} |\nabla u_j|^2 \, dx = \int_{\mathbb{R}^N} |\nabla u'_j|^2 \, dx + \int_{\mathbb{R}^N} |\nabla u''_j|^2 \, dx + 2 \int_{\mathbb{R}^N} \nabla u'_j \cdot \nabla u''_j \, dx.$$

Then we have

$$\int_{\mathbb{R}^N} \nabla u'_j \cdot \nabla u''_j \, dx = \int_{\mathbb{R}^N} \left(u_j \nabla \psi_j + \psi_j \nabla u_j \right) \cdot \left((1 - \psi_j) \nabla u_j - u_j \nabla \psi_j \right) dx$$

$$= \int_{\mathbb{R}^N} u_j (1 - \psi_j) \nabla \psi_j \cdot \nabla u_j \, dx - \int_{\mathbb{R}^N} u_j^2 |\nabla \psi_j|^2 \, dx$$

$$+ \int_{\mathbb{R}^N} \psi_j (1 - \psi_j) |\nabla u_j|^2 \, dx - \int_{\mathbb{R}^N} \psi_j u_j \nabla u_j \cdot \nabla \psi_j \, dx$$

$$\geq \int_{\mathbb{R}^N} u_j (1 - \psi_j) \nabla \psi_j \cdot \nabla u_j \, dx - \int_{\mathbb{R}^N} u_j^2 |\nabla \psi_j|^2 \, dx$$

$$- \int_{\mathbb{R}^N} \psi_j u_j \nabla u_j \cdot \nabla \psi_j \, dx.$$

We also have

$$\int_{\mathbb{R}^N} u_j^2 |\nabla \psi_j|^2 \, dx \leq C \int_{C_j} u_j^2 \, dx \leq \frac{2C}{j} = o(1),$$

and

$$\left| \int_{\mathbb{R}^N} \psi_j u_j \nabla u_j \cdot \nabla \psi_j \, dx \right| \leq \int_{\mathbb{R}^N} \psi_j u_j |\nabla u_j| |\nabla \psi_j| \, dx$$

$$\leq \left(\int_{\mathbb{R}^N} |\nabla u_j|^2 \, dx \right)^{1/2} \left(\int_{\mathbb{R}^N} u_j^2 |\nabla \psi_j|^2 \, dx \right)^{1/2}$$

$$\leq M \left(\frac{2C}{j} \right)^{1/2} = o(1),$$

where $M = \sup_j (\int_{\mathbb{R}^N} |\nabla u_j|^2 \, dx)^{1/2}$.

In the same way we obtain

$$\left| \int_{\mathbb{R}^N} (1 - \psi_j) u_j \nabla u_j \cdot \nabla \psi_j \, dx \right| \leq o(1),$$

and putting all the estimates together we deduce that

$$\int_{\mathbb{R}^N} |\nabla u_j|^2 \, dx \geq \int_{\mathbb{R}^N} |\nabla u'_j|^2 \, dx + \int_{\mathbb{R}^N} |\nabla u''_j|^2 \, dx + o(1),$$

and finally,

$$\|u_j\|^2 \geq \|u'_j\|^2 + \|u''_j\|^2 + o(1). \qquad \square$$

Lemma 3.3.10 *There results*

$$\left(\frac{1}{2} - \frac{1}{r} \right) \|u\|^2 + \left(\frac{1}{r} - \frac{1}{p} \right) |u|_p^p \leq m.$$

Proof Since $u_k \rightharpoonup u$ in $H^1(\mathbb{R}^N)$ and in $L^p(\mathbb{R}^N)$,

$$\left(\frac{1}{2} - \frac{1}{r}\right)\|u\|^2 + \left(\frac{1}{r} - \frac{1}{p}\right)|u|_p^p$$
$$\leq \left(\frac{1}{2} - \frac{1}{r}\right)\liminf_k \|u_k\|^2 + \left(\frac{1}{r} - \frac{1}{p}\right)\liminf_k |u_k|_p^p$$
$$\leq \liminf_k I(u_k) = m. \qquad \square$$

We now prove that $u \in \mathcal{N}$. We know that $u \neq 0$, so we just have to check that $\|u\|^2 = |u|_p^p + \lambda|u|_r^r$. First we rule out the easy inequality.

Lemma 3.3.11 *It cannot be*

$$\|u\|^2 < |u|_p^p + \lambda|u|_r^r.$$

Proof If $\|u\|^2 < |u|_p^p + \lambda|u|_r^r$, then, by usual arguments, there is some $t \in (0, 1)$ such that $tu \in \mathcal{N}$, so that

$$m \leq I(tu) = t^2\left(\frac{1}{2} - \frac{1}{r}\right)\|u\|^2 + t^p\left(\frac{1}{r} - \frac{1}{p}\right)|u|_p^p$$
$$< \left(\frac{1}{2} - \frac{1}{r}\right)\|u\|^2 + \left(\frac{1}{r} - \frac{1}{p}\right)|u|_p^p \leq m,$$

a contradiction. $\qquad \square$

Then, with some more effort, the opposite case.

Lemma 3.3.12 *It cannot be*

$$\|u\|^2 > |u|_p^p + \lambda|u|_r^r.$$

Proof Using Lemma 3.3.9 we can write

$$\|u_j'\|^2 + \|u_j''\|^2 \leq \|u_j\|^2 + o(1) = |u_j|_p^p + \lambda|u_j|_r^r + o(1)$$
$$= |u_j'|_p^p + \lambda|u_j'|_r^r + |u_j''|_p^p + \lambda|u_j''|_r^r + o(1). \qquad (3.10)$$

Assume for contradiction that $\|u\|^2 > |u|_p^p + \lambda|u|_r^r$ and pick $\delta > 0$ such that

$$\|u\|^2 > |u|_p^p + \lambda|u|_q^q + \delta.$$

Since $u_j' \rightharpoonup u$ in $H^1(\mathbb{R}^N)$,

$$\liminf_j \|u_j'\|^2 \geq \|u\|^2,$$

while $u_j' \to u$ in $L^p(\mathbb{R}^N)$ and in $L^r(\mathbb{R}^N)$. From this we deduce that, for large j's,

$$\|u_j'\|^2 > |u_j'|_p^p + \lambda|u_j'|_r^r + \delta_1,$$

for some $\delta_1 \in (0, \delta)$. Together with (3.10), this implies

$$\|u_j''\|^2 \leq |u_j''|_p^p + \lambda|u_j''|_r^r - \delta_2,$$

for large j's and for some $\delta_2 \in (0, \delta_1)$. Recalling that $u_j'' \rightharpoonup 0$ in $H^1(\mathbb{R}^N)$ and arguing as in Lemma 3.3.8, we obtain

$$\lim_j \|u_j''\|^2 = \lim_j \|u_j''\|_\alpha^2,$$

so that

$$\|u_j''\|_\alpha^2 \leq |u_j''|_p^p + \lambda |u_j''|_r^r - \delta_3,$$

for large j's and for some $\delta_3 \in (0, \delta_2)$. Hence, by usual arguments, we can find $s_j \in (0, 1)$ such that $s_j u_j'' \in \mathcal{N}_\alpha$. So we obtain

$$m_\alpha \leq I_\alpha(s_j u_j'') = s_j^2 \left(\frac{1}{2} - \frac{1}{r}\right) \|u_j''\|_\alpha^2 + s_j^p \left(\frac{1}{r} - \frac{1}{p}\right) |u_j''|_p^p$$

$$\leq \left(\frac{1}{2} - \frac{1}{r}\right) \|u_j''\|^2 + \left(\frac{1}{r} - \frac{1}{p}\right) |u_j''|_p^p + o(1)$$

$$\leq \left(\frac{1}{2} - \frac{1}{r}\right) \left(\|u_j''\|^2 + \|u_j'\|^2\right) + \left(\frac{1}{r} - \frac{1}{p}\right) \left(|u_j''|_p^p + |u_j'|_p^p\right) + o(1)$$

$$\leq \left(\frac{1}{2} - \frac{1}{r}\right) \|u_j\|^2 + \left(\frac{1}{r} - \frac{1}{p}\right) |u_j|_p^p + o(1) = I(u_j) + o(1).$$

Passing to the limit this yields

$$m_\alpha \leq m,$$

a contradiction. □

End of the proof of Theorem 3.3.2 We can now conclude easily the proof of the main result. The previous lemmas say that in any case the weak limit u satisfies $u \neq 0$, $u \geq 0$ and

$$\|u\|^2 = |u|_p^p + \lambda |u|_r^r.$$

Therefore $u \in \mathcal{N}$ and also

$$I(u) = \left(\frac{1}{2} - \frac{1}{r}\right) \|u\|^2 + \left(\frac{1}{r} - \frac{1}{p}\right) |u|_p^p \leq m,$$

so that $I(u) = m$ and u is a minimum point for I on $\mathcal{N}$. It is then easy, arguing as in the previous sections, to prove that u is a critical point for I on $H^1(\mathbb{R}^N)$. □

3.4 Problems with Critical Exponent

We discuss in this section some examples of problems where the nonlinearity has *critical* growth. This is one of the most dramatic cases of loss of compactness and has been studied intensively in the last 25 years, starting with the pioneering paper [14].

We present here only simple cases, with the aim of showing some example of the techniques involved.

3.4.1 The Prototype Problem

We consider in $\mathbb{R}^N$, with $N \geq 3$, the equation

$$-\Delta u = |u|^{2^*-2}u. \qquad (3.11)$$

Formally, solutions of (3.11) should arise as critical points of the functional

$$I(u) = \frac{1}{2}\int_{\mathbb{R}^N} |\nabla u|^2\, dx - \frac{1}{2^*}\int_{\mathbb{R}^N} |u|^{2^*}\, dx. \qquad (3.12)$$

It is not convenient to think of I as defined on $H^1(\mathbb{R}^N)$, since in this space the first integral in I is not (the square of) a norm, and this poses serious difficulties.

We will work in the space $D^{1,2}(\mathbb{R}^N)$ that we introduced in Sect. 1.2. The problem we are dealing with then becomes

$$\begin{cases} -\Delta u = |u|^{2^*-2}u, \\ u \in D^{1,2}(\mathbb{R}^N). \end{cases} \qquad (3.13)$$

We know that $D^{1,2}(\mathbb{R}^N) \subset L^{2^*}(\mathbb{R}^N)$, with continuous embedding, but the embedding is *not* compact, not even in a local sense, as we now make precise.

Clearly, since we are working in $\mathbb{R}^N$, there is a lack of compactness due to the invariance under translations, as we discussed in earlier sections. However in this case there is an even more serious problem, that arises from the *invariance under scalings*, as follows. Let $v \in C_0^\infty(\mathbb{R}^N)$ (hence $v \in D^{1,2}(\mathbb{R}^N)$) be a fixed function, and set, for $\lambda > 0$,

$$v_\lambda(x) = \lambda^{\frac{N-2}{2}} v(\lambda x).$$

Then it is an exercise to check that

$$\int_{\mathbb{R}^N} |\nabla v_\lambda|^2\, dx = \int_{\mathbb{R}^N} |\nabla v|^2\, dx \quad \text{for every } \lambda > 0$$

and

$$v_\lambda \rightharpoonup 0 \quad \text{in } D^{1,2}(\mathbb{R}^N) \quad \text{for } \lambda \to 0 \text{ or } \lambda \to +\infty.$$

However also

$$\int_{\mathbb{R}^N} |v_\lambda|^{2^*}\, dx = \int_{\mathbb{R}^N} |v|^{2^*}\, dx \quad \text{for every } \lambda > 0, \qquad (3.14)$$

so that $\{v_\lambda\}_\lambda$ is not precompact in $L^{2^*}(\mathbb{R}^N)$. This happens exactly because the exponent is 2^*.

With respect to the informal discussion on the loss of compactness carried out in the preceding section, we have to face here new ways in which this phenomenon takes place. Indeed, assuming to fix ideas that $v(0) \neq 0$, we see that when $\lambda \to 0$ the functions $v_\lambda \in C_0^\infty(\mathbb{R}^N)$ tend to zero uniformly, while $\int_{\mathbb{R}^N} |u|^{2^*}\, dx$ is kept constant (see (3.14)). One can say that the "mass" spreads everywhere. On the contrary, if $\lambda \to \infty$, then $v_\lambda(x) \to 0$ for every $x \neq 0$, while $v_\lambda(0) \to \infty$, and again (3.14) holds; this is the famous loss of compactness *by concentration*, that one meets in every problem with critical growth, even on bounded domains.

The hard part in the proof of the main result of this section will be devoted to a careful analysis of minimizing sequences to understand the consequences of spreading or concentration of mass.

Anyway, the first thing to do if one wants to solve (3.13), is to get rid of the invariance under translations. This can be done by symmetrization, as in Sect. 3.1.

We recall that we have defined

$$D_r = \left\{ u \in D^{1,2}(\mathbb{R}^N) \mid u \text{ is radial} \right\}.$$

We are going to prove the following theorem.

Theorem 3.4.1 *There exists a nonnegative and nontrivial $u \in D_r$ that solves weakly Problem* (3.13), *namely such that*

$$\int_{\mathbb{R}^N} \nabla u \cdot \nabla v \, dx = \int_{\mathbb{R}^N} u^{2^*-1} v \, dx \qquad (3.15)$$

for every v in $D^{1,2}(\mathbb{R}^N)$.

Remark 3.4.2 It can be proved, with essentially the same arguments as in Example 1.3.21, that the functional

$$u \mapsto \int_{\mathbb{R}^N} |u|^{2^*} \, dx$$

(and hence the functional I) is differentiable on $D^{1,2}(\mathbb{R}^N)$. This does not follow directly from Example 1.3.21, because the H^1 norm is larger that the $D^{1,2}$ norm. In other words, quantities that are $o(\|v\|_{H^1})$ are not, generally, $o(\|v\|_{D^{1,2}})$. A slightly more general case is reported in Exercise 4.

In the proof of Theorem 3.4.1 we will need the pointwise estimate that we have proved in Lemma 3.1.2.

Now we can start with the main argument. We want to minimize the functional I on the unit sphere of $L^{2^*}(\mathbb{R}^N)$. This amounts to setting

$$S = \inf \left\{ \|u\|^2 \,\middle|\, u \in D^{1,2}(\mathbb{R}^N), \int_{\mathbb{R}^N} |u|^{2^*} \, dx = 1 \right\}. \qquad (3.16)$$

Remark 3.4.3 The value S is known as the *best Sobolev constant*. It is the largest positive constant S such that

$$S|u|_{2^*}^2 \le \|u\|^2 \quad \text{for every } u \in D^{1,2}(\mathbb{R}^N).$$

Thanks to the Sobolev inequalities, we know that $S > 0$. We will show that the infimum S is attained by some $u \in D^{1,2}(\mathbb{R}^N)$.

We begin to carry out the analysis of minimizing sequences by selecting one that enjoys some special properties, as we show in the next three lemmas. To begin with, we need a minimizing sequence made by *radial* functions.

Lemma 3.4.4 *There exists a sequence* $\{v_k\}_k \subset D_r$ *such that* $v_k \geq 0$,

$$|v_k|_{2^*} = 1 \quad and \quad \|v_k\|^2 \to S.$$

Proof Let $\{w_k\}_k \subset D^{1,2}(\mathbb{R}^N)$ be a minimizing sequence for S; of course it is not restrictive to assume that $w_k \geq 0$. Define $v_k = w_k^* \geq 0$ (see Theorem 3.1.5). Then

$$v_k \in D_r, \quad |v_k|_{2^*} = |w_k|_{2^*} = 1 \quad and \quad \|v_k\|^2 \leq \|w_k\|^2,$$

so that $\{v_k\}_k \subset D_r$ is a minimizing sequence for S. $\square$

Lemma 3.4.5 *There exist a minimizing sequence* $\{u_k\}_k \subset D_r$ *for* S *and a function* $u \in D_r$ *with the following properties:*

- $u_k \geq 0, u \geq 0$.
- $\int_{|x|<1} |u_k|^{2^*} \, dx = \int_{|x|\geq 1} |u_k|^{2^*} \, dx = \frac{1}{2}$.
- $u_k \rightharpoonup u$ *in* $D^{1,2}(\mathbb{R}^N)$ *and in* $L^{2^*}(\mathbb{R}^N)$, *and* $u_k(x) \to u(x)$ *a.e. in* $\mathbb{R}^N$.

Proof Let $\{v_k\}_k \subset D_r$ be a minimizing sequence for S, with $v_k \geq 0$. Since

$$\int_{\mathbb{R}^N} |v_k|^{2^*} \, dx = 1,$$

there exists $R_k > 0$ such that

$$\int_{|x|<R_k} |v_k|^{2^*} \, dx = \int_{|x|\geq R_k} |v_k|^{2^*} \, dx = \frac{1}{2}.$$

If we define

$$u_k(x) = R_k^{\frac{N-2}{2}} v_k(R_k x),$$

it is immediate to check that $u_k \in D_r$ and that $\|u_k\| = \|v_k\|$ and $|u_k|_{2^*} = |v_k|_{2^*}$. This means that $\{u_k\}_k$ is a minimizing sequence too. Of course $u_k \geq 0$, and a simple change of variables shows that

$$\int_{|x|<1} |u_k|^{2^*} \, dx = \int_{|x|\geq 1} |u_k|^{2^*} \, dx = \frac{1}{2}.$$

Since $\{u_k\}_k$ is bounded in $D^{1,2}(\mathbb{R}^N)$, up to subsequences, $u_k \rightharpoonup u$ in $D^{1,2}(\mathbb{R}^N)$ and in $L^{2^*}(\mathbb{R}^N)$. Moreover, for every ball B, the sequence u_k is bounded in $H^1(B)$, so that, again up to subsequences, it converges strongly in $L^2(B)$ and also satisfies $u_k(x) \to u(x)$ almost everywhere in $\mathbb{R}^N$. The proof is complete. $\square$

Next, we force the sequence u_k to have the technical property stated in the following lemma.

Lemma 3.4.6 *There exists a minimizing sequence* $\{u_k\}_k$ *for* S *that, in addition to those of the previous lemma, satisfies also the following properties: for every* $j \in \mathbb{N}$, $j \geq 1$, *the sequences*

$$\left\{ \int_{|x|<1/j} |\nabla u_k|^2 \, dx \right\}_k \quad and \quad \left\{ \int_{|x|>j} |\nabla u_k|^2 \, dx \right\}_k$$

admit finite limits as $k \to \infty$.

Proof Let $\{u_k\}_k$ be the sequence defined in the previous lemma. Since the numerical sequences

$$\int_{|x|<1} |\nabla u_k|^2 \, dx \quad \text{and} \quad \int_{|x|>1} |\nabla u_k|^2 \, dx$$

are bounded, there exists a subsequence $\{u_{k,1}\}_k$ such that the sequences

$$\int_{|x|<1} |\nabla u_{k,1}|^2 \, dx \quad \text{and} \quad \int_{|x|>1} |\nabla u_{k,1}|^2 \, dx$$

admit finite limits. From $\{u_{k,1}\}_k$ we can extract a new subsequence $\{u_{k,2}\}_k$ such that the sequences

$$\int_{|x|<1/2} |\nabla u_{k,2}|^2 \, dx \quad \text{and} \quad \int_{|x|>2} |\nabla u_{k,2}|^2 \, dx$$

admit finite limits. Proceeding in this way, for every $j \geq 2$ we have a sequence $\{u_{k,j}\}_k$ such that the sequences

$$\int_{|x|<1/j} |\nabla u_{k,j}|^2 \, dx \quad \text{and} \quad \int_{|x|>j} |\nabla u_{k,j}|^2 \, dx$$

admit finite limits as $k \to \infty$ and $\{u_{k,j}\}_k$ is a subsequence of $\{u_{k,j-1}\}_k$. Then we relabel as u_k the "diagonal", namely the sequence $u_k = u_{k,k}$. This is a minimizing sequence that satisfies the conditions of the previous lemma, because it has been extracted from the original minimizing sequence. Moreover, for every given j, except the first j terms, the sequence is a subsequence of $\{u_{k,j}\}_k$, and therefore the numerical sequences

$$\int_{|x|<1/j} |\nabla u_k|^2 \, dx \quad \text{and} \quad \int_{|x|>j} |\nabla u_k|^2 \, dx$$

are subsequences of a convergent sequence, and therefore they are convergent too. $\square$

In the next lemma we introduce some quantities that we will use in the main argument. The reader can recognize the effort in trying to gain information on the spreading or on the concentration of u_k, as $k \to \infty$.

Lemma 3.4.7 *The following limits exist and are finite:*

1. $v_0 = \lim_{R \to 0^+} (\limsup_k \int_{|x|<R} |u_k|^{2^*} \, dx)$,
   $v_\infty = \lim_{R \to +\infty} (\limsup_k \int_{|x|\geq R} |u_k|^{2^*} \, dx)$.
2. $\mu_0 = \lim_{R \to 0^+} (\limsup_k \int_{|x|<R} |\nabla u_k|^2 \, dx)$,
   $\mu_\infty = \lim_{R \to +\infty} (\limsup_k \int_{|x|\geq R} |\nabla u_k|^2 \, dx)$.

Moreover

3. $v_0, v_\infty \in [0, 1/2]$.

Proof Defining the functions

$$f(R) = \limsup_k \int_{|x|<R} |u_k|^{2^*} dx \quad \text{and} \quad g(R) = \limsup_k \int_{|x|\geq R} |u_k|^{2^*} dx,$$

it is immediate to see that f is bounded and nondecreasing, while g is bounded and nonincreasing. Therefore

$$\lim_{R\to 0^+} f(R) = \nu_0 \quad \text{and} \quad \lim_{R\to +\infty} g(R) = \nu_\infty$$

exist and are finite (and nonnegative).

The same argument works for μ_0 and μ_∞. The fact that $\nu_0 \leq 1/2$ and $\nu_\infty \leq 1/2$ comes from

$$\int_{|x|<1} |u_k|^{2^*} dx = \int_{|x|\geq 1} |u_k|^{2^*} dx = \frac{1}{2}. \qquad \square$$

The following lemmas establish some inequalities among the numbers introduced in the above result that will be essential in the proof of Theorem 3.4.1. Loosely speaking, the next result establishes that as $k \to \infty$, the original mass of u_k in not less than the sum of the contributions of the weak limit, of the part that spreads (ν_∞) and of the part that concentrates (ν_0).

Lemma 3.4.8 *Let u be the weak limit of the minimizing sequence u_k constructed above. Then*

$$1 \leq \int_{\mathbb{R}^N} |u|^{2^*} dx + \nu_0 + \nu_\infty.$$

Proof For every $k \in \mathbb{N}$ and for every $R > 1$ we have

$$1 = \int_{\mathbb{R}^N} |u_k|^{2^*} dx = \int_{|x|<1/R} |u_k|^{2^*} dx + \int_{1/R<|x|<R} |u_k|^{2^*} dx + \int_{|x|\geq R} |u_k|^{2^*} dx.$$

In the bounded open set

$$\{x \in \mathbb{R}^N \mid 1/R < |x| < R\}$$

we have $|u_k(x)|^{2^*} \to |u(x)|^{2^*}$ almost everywhere and moreover

$$|u_k(x)|^{2^*} \leq CR^N,$$

by Lemma 3.1.2. Then, by dominated convergence,

$$\lim_k \int_{1/R<|x|<R} |u_k|^{2^*} dx = \int_{1/R<|x|<R} |u|^{2^*} dx.$$

Then we have

$$1 = \limsup_k \left(\int_{|x|<1/R} |u_k|^{2^*} dx + \int_{1/R<|x|<R} |u_k|^{2^*} dx + \int_{|x|\geq R} |u_k|^{2^*} dx \right)$$

$$\leq \int_{1/R<|x|<R} |u|^{2^*} dx + \limsup_k \int_{|x|<1/R} |u_k|^{2^*} dx + \limsup_k \int_{|x|\geq R} |u_k|^{2^*} dx.$$

As $R \to +\infty$ we obtain

$$1 \leq \int_{\mathbb{R}^N} |u|^{2^*} dx + \nu_0 + \nu_\infty. \qquad \square$$

A similar behavior (but with the opposite inequality) holds also for the $D^{1,2}$ norm.

Lemma 3.4.9 *Let u be the weak limit of the minimizing sequence u_k constructed above. Then*

$$S \geq \int_{\mathbb{R}^N} |\nabla u|^2 dx + \mu_0 + \mu_\infty.$$

Proof Setting

$$\varepsilon_k = \|u_k\|^2 - S,$$

for every k and every j there results

$$S + \varepsilon_k = \int_{\mathbb{R}^N} |\nabla u_k|^2 dx$$

$$= \int_{|x|<1/j} |\nabla u_k|^2 dx + \int_{1/j<|x|<j} |\nabla u_k|^2 dx + \int_{|x|>j} |\nabla u_k|^2 dx. \quad (3.17)$$

The functional

$$J(u) = \int_{1/j<|x|<j} |\nabla u|^2 dx$$

is continuous and convex in $D^{1,2}(\mathbb{R}^N)$, so that it is weakly lower semicontinuous. Recalling that $\varepsilon_k \to 0$ and that the two numerical sequences

$$\left\{ \int_{|x|<1/j} |\nabla u_k|^2 dx \right\}_k \quad \text{and} \quad \left\{ \int_{|x|>j} |\nabla u_k|^2 dx \right\}_k$$

admit limits, if in (3.17) we take the lim inf with respect to k, we obtain

$$S = \liminf_k \left(\int_{|x|<1/j} |\nabla u_k|^2 dx + \int_{1/j<|x|<j} |\nabla u_k|^2 dx + \int_{|x|>j} |\nabla u_k|^2 dx \right)$$

$$= \liminf_k \int_{1/j<|x|<j} |\nabla u_k|^2 dx + \lim_k \int_{|x|<1/j} |\nabla u_k|^2 dx + \lim_k \int_{|x|>j} |\nabla u_k|^2 dx$$

$$\geq \int_{1/j<|x|<j} |\nabla u|^2 dx + \lim_k \int_{|x|<1/j} |\nabla u_k|^2 dx + \lim_k \int_{|x|>j} |\nabla u_k|^2 dx.$$

Letting now $j \to \infty$ we see that

$$S \geq \int_{\mathbb{R}^N} |\nabla u|^2 dx + \mu_0 + \mu_\infty. \qquad \square$$

From the definition of S one has, for every $v \in D^{1,2}(\mathbb{R}^N)$,

$$S \left(\int_{\mathbb{R}^N} |v|^{2^*} dx \right)^{2/2^*} \leq \int_{\mathbb{R}^N} |\nabla v|^2 dx. \qquad (3.18)$$

The next lemma is of fundamental importance. In some sense it says that the inequality (3.18) holds also for the "components" of u_k that spread or concentrate.

Lemma 3.4.10 *There results*

$$Sv_0^{2/2^*} \leq \mu_0 \quad and \quad Sv_\infty^{2/2^*} \leq \mu_\infty.$$

Proof Let $\varphi : \mathbb{R} \to \mathbb{R}$ be a C^∞ function such that $\varphi(t) = 1$ if $t \in [0, 1]$ and $\varphi(t) = 0$ if $t \geq 2$. For every $R > 0$ and every $x \in \mathbb{R}^N$, set

$$\varphi_R(x) = \varphi\left(\frac{|x|}{R}\right).$$

Then $\varphi_R \in C_0^\infty(\mathbb{R}^N)$,

$$\begin{cases} \varphi_R(x) = 1 & \text{if } |x| \leq R, \\ \varphi_R(x) = 0 & \text{if } |x| \geq 2R, \\ |\nabla\varphi_R(x)| \leq C/R & \text{for every } x \in \mathbb{R}^N \end{cases}$$

for some C independent of x and R.

For every $v \in D^{1,2}(\mathbb{R}^N)$, the function $\varphi_R v$ is also in $D^{1,2}(\mathbb{R}^N)$. Indeed $\varphi_R v$ belongs to $L^{2^*}(\mathbb{R}^N)$, and we only have to check that its gradient is in L^2. But

$$\nabla(\varphi_R v) = v\nabla\varphi_R + \varphi_R\nabla v,$$

and obviously $|\varphi_R \nabla v| \in L^2(\mathbb{R}^N)$. On the other hand,

$$\int_{\mathbb{R}^N} |v|^2 |\nabla\varphi_R|^2 \, dx = \int_{R<|x|<2R} |v|^2 |\nabla\varphi_R|^2 \, dx$$

$$\leq \left(\int_{R<|x|<2R} |v|^{2^*}\right)^{\frac{N-2}{N}} \left(\int_{R<|x|<2R} |\nabla\varphi_R|^N \, dx\right)^{2/N}$$

$$\leq \left(\int_{\mathbb{R}^N} |v|^{2^*}\right)^{\frac{N-2}{N}} \left(\frac{C}{R^N} \int_{R<|x|<2R} dx\right)^{2/N} \leq C|v|_{2^*}^2 < +\infty.$$

Thus, the gradient of $\varphi_R v$ is the sum of two L^2 functions.

We now consider, for every k and for every R, the function $\varphi_R u_k \in D^{1,2}(\mathbb{R}^N)$. Certainly,

$$S\left(\int_{\mathbb{R}^N} |\varphi_R u_k|^{2^*} \, dx\right)^{2/2^*} \leq \int_{\mathbb{R}^N} |\nabla(\varphi_R u_k)|^2 \, dx. \tag{3.19}$$

We analyze what happens to this inequality when we compute some limits. As far as the left-hand side is concerned, we have

$$\int_{|x|<R} |u_k|^{2^*} \, dx \leq \int_{\mathbb{R}^N} |\varphi_R u_k|^{2^*} \, dx \leq \int_{|x|<2R} |u_k|^{2^*} \, dx,$$

so that

$$\limsup_k \int_{|x|<R} |u_k|^{2^*} \, dx \leq \limsup_k \int_{\mathbb{R}^N} |\varphi_R u_k|^{2^*} \, dx \leq \limsup_k \int_{|x|<2R} |u_k|^{2^*} \, dx.$$

Now the first and the last term in this inequality tend to v_0, when R tends to zero (Lemma 3.4.7), and therefore so does the central term, namely

$$\lim_{R \to 0} \left(\limsup_k \int_{\mathbb{R}^N} |\varphi_R u_k|^{2^*} \, dx \right) = v_0.$$

We turn to the study of the right-hand side of (3.19). First of all,

$$|\nabla(\varphi_R u_k)|^2 = u_k^2 |\nabla \varphi_R|^2 + \varphi_R^2 |\nabla u_k|^2 + 2 u_k \varphi_R \nabla \varphi_R \cdot \nabla u_k.$$

Repeating the above argument, we deduce that

$$\int_{\mathbb{R}^N} |u_k|^2 |\nabla \varphi_R|^2 \, dx \le C \left(\int_{R < |x| < 2R} |u_k|^{2^*} \, dx \right)^{\frac{N-2}{N}}.$$

Moreover, working as in Lemma 3.4.8, we see that

$$\lim_k \int_{R < |x| < 2R} |u_k|^{2^*} \, dx = \int_{R < |x| < 2R} |u|^{2^*} \, dx$$

so that

$$\lim_{R \to 0} \left(\lim_k \int_{R < |x| < 2R} |u_k|^{2^*} \, dx \right) = \lim_{R \to 0} \int_{R < |x| < 2R} |u|^{2^*} \, dx = 0,$$

which yields

$$\lim_{R \to 0} \left(\lim_k \int_{\mathbb{R}^N} |u_k|^2 |\nabla \varphi_R|^2 \, dx \right) = 0.$$

From this it also easy to see that

$$\lim_{R \to 0} \left(\lim_k \int_{\mathbb{R}^N} u_k \varphi_R \nabla \varphi_R \cdot \nabla u_k \, dx \right) = 0.$$

Finally, since

$$\int_{|x| < R} |\nabla u_k|^2 \, dx \le \int_{\mathbb{R}^N} |\varphi_R \nabla u_k|^2 \, dx \le \int_{|x| < 2R} |\nabla u_k|^2 \, dx,$$

arguing as above we obtain

$$\lim_{R \to 0} \left(\limsup_k \int_{\mathbb{R}^N} |\varphi_R \nabla u_k|^2 \, dx \right) = \mu_0.$$

Summing up, from (3.19), taking first the lim sup over k, and then the limit as $R \to 0$, we get to

$$S v_0^{2/2^*} \le \mu_0,$$

the first inequality we wanted to prove.

The other inequality can be deduced in a similar way. Defining $\psi : \mathbb{R} \to \mathbb{R}$ to be a C^∞ function such that $\psi(t) = 0$ if $t \in [0, 1]$ and $\psi(t) = 1$ if $t \ge 2$, for every $R > 0$ and every $x \in \mathbb{R}^N$ we set

$$\psi_R(x) = \psi \left(\frac{|x|}{R} \right).$$

Again, it is immediate to check that $\psi_R \in C^\infty(\mathbb{R}^N)$,

$$
\begin{cases}
\psi_R(x) = 0 & \text{if } |x| \le R, \\
\psi_R(x) = 1 & \text{if } |x| \ge 2R, \\
|\nabla \psi_R(x)| \le C/R & \text{for every } x \in \mathbb{R}^N
\end{cases}
$$

for some C independent of x and R. As before, $\psi_R v \in D^{1,2}(\mathbb{R}^N)$ if $v \in D^{1,2}(\mathbb{R}^N)$.

Starting from the inequality

$$
S\left(\int_{\mathbb{R}^N} |\psi_R u_k|^{2^*} dx \right)^{2/2^*} \le \int_{\mathbb{R}^N} |\nabla(\psi_R u_k)|^2 dx
$$

and arguing as in the first part of the proof, we obtain

$$
\lim_{R \to +\infty} \left(\limsup_k \int_{\mathbb{R}^N} |\psi_R u_k|^{2^*} dx \right) = v_\infty
$$

and

$$
\lim_{R \to +\infty} \left(\limsup_k \int_{\mathbb{R}^N} |\psi_R \nabla u_k|^2 dx \right) = \mu_\infty,
$$

so that also the inequality

$$
S v_\infty^{2/2^*} \le \mu_\infty
$$

is established.   □

The next lemma is the final step in the proof of Theorem 3.4.1.

Lemma 3.4.11 *There results*

$$
v_0 = v_\infty = 0 \quad and \quad \int_{\mathbb{R}^N} |u|^{2^*} dx = 1.
$$

Proof We know that $v_0, v_\infty \in [0, 1/2]$ and, by weak lower semicontinuity,

$$
\int_{\mathbb{R}^N} |u|^{2^*} dx \in [0, 1].
$$

Since $2^* > 2$, for every $a \in [0, 1]$ we have $a^{2/2^*} \ge a$, and equality holds if and only if $a = 0$ or $a = 1$.

Putting together the inequalities proved in the preceding lemmas, we have

$$
\begin{aligned}
S &\ge \int_{\mathbb{R}^N} |\nabla u|^2 dx + \mu_0 + \mu_\infty \\
&\ge S\left[\left(\int_{\mathbb{R}^N} |u|^{2^*} dx \right)^{2/2^*} + v_0^{2/2^*} + v_\infty^{2/2^*} \right] \\
&\ge S\left[\int_{\mathbb{R}^N} |u|^{2^*} dx + v_0 + v_\infty \right] \ge S.
\end{aligned}
$$

Thus equality must hold everywhere, and in particular

$$\left(\int_{\mathbb{R}^N} |u|^{2^*} dx\right)^{2/2^*} + v_0^{2/2^*} + v_\infty^{2/2^*} = \int_{\mathbb{R}^N} |u|^{2^*} dx + v_0 + v_\infty.$$

But then, each of the numbers $\int_{\mathbb{R}^N} |u|^{2^*} dx$, v_0 and v_∞ must either be 0 or 1. As $v_0, v_\infty \in [0, 1/2]$, necessarily $v_0 = v_\infty = 0$.

Since

$$\int_{\mathbb{R}^N} |u|^{2^*} dx + v_0 + v_\infty \ge 1,$$

we deduce that $\int_{\mathbb{R}^N} |u|^{2^*} dx \ge 1$ and hence $\int_{\mathbb{R}^N} |u|^{2^*} dx = 1$. $\qquad\square$

End of the proof of Theorem 3.4.1 We have proved so far that there exists a minimizing sequence for S that admits a weak limit u satisfying

$$\int_{\mathbb{R}^N} |u|^{2^*} dx = 1. \tag{3.20}$$

By weak lower semicontinuity, we also have $\|u\|^2 \le S$, and by the definition of S and (3.20), it must be $\|u\|^2 = S$. Thus u is a minimizer.

Now the fact that a multiple of u weakly solves (3.13) follows from standard arguments already used in Chap. 2. The proof of Theorem 3.4.1 is now complete. $\square$

Remark 3.4.12 The solutions of (3.13) that we have found are explicitly known. Elementary computations show that a solution is given, for a suitable constant $c > 0$, by

$$U(x) = \frac{c}{\left(1 + |x|^2\right)^{\frac{N-2}{2}}},$$

and it can be proved (it is not easy at all) that this function indeed solves the minimization problem (3.16).

Actually *all* the solutions of (3.16) are known: they are the family of functions

$$U_{\lambda,y}(x) = \lambda^{\frac{N-2}{2}} U(\lambda(x - y))$$

for each $y \in \mathbb{R}^N$ and each $\lambda > 0$. Note that they are generated by the function u above by translations and scalings (see e.g. [29]).

3.4.2 A Problem with a Radial Coefficient

We now use the results of the previous section to deal with a problem that depends on a radial coefficient. It is a simple variant of (3.13).

We consider

$$\begin{cases} -\Delta u = \beta(|x|)|u|^{2^*-2}u, \\ u \in D^{1,2}(\mathbb{R}^N), \end{cases} \tag{3.21}$$

where β satisfies the following assumptions

(b$_1$) $\beta : [0, +\infty) \to [0, +\infty)$ is continuous.

(b$_2$) $\beta(r) > \beta(0)$ for all $r > 0$ and $\lim_{r \to +\infty} \beta(r) = \beta(0)$.

For example, the functions $\beta(r) = C + \frac{r}{1+r^2}$ or $\beta(r) = C + re^{-r}$ satisfy the above assumptions for every $C \geq 0$. Hypothesis **(b$_2$)** can be weakened, see Remark 3.4.22 below.

We will prove the following result.

Theorem 3.4.13 *Assume that* **(b$_2$)** *and* **(b$_2$)** *hold. Then there exists a nonnegative and nontrivial $u \in D_r$ that solves weakly Problem* (3.21), *namely such that*

$$\int_{\mathbb{R}^N} \nabla u \cdot \nabla v \, dx = \int_{\mathbb{R}^N} \beta(|x|) u^{2^*-1} v \, dx \qquad (3.22)$$

for every v in $D^{1,2}(\mathbb{R}^N)$.

To prove this result we cannot use the symmetrization technique as in the previous section, that is, we cannot apply Theorem 3.1.5. The reason is the fact that, if $v \in D^{1,2}(\mathbb{R}^N)$ and $v = u^*$ (see Theorem 3.1.5), it is true that $\|u\|^2 \leq \|v\|^2$ but the equality

$$\int_{\mathbb{R}^N} \beta(|x|) u^{2^*} \, dx = \int_{\mathbb{R}^N} \beta(|x|) v^{2^*} \, dx$$

does not hold.

Hence we adopt a different strategy: we will study the variational problem not in the space $D^{1,2}(\mathbb{R}^N)$ but *directly* in the space D_r. Of course this is possible because the problem is invariant under rotations.

We begin by introducing the following definition. For every positive number $b > 0$ we set

$$S_b = \inf \left\{ \|u\|^2 \;\middle|\; u \in D^{1,2}(\mathbb{R}^N), \int_{\mathbb{R}^N} |u|^{2^*} \, dx = b \right\}.$$

Notice that S_1 coincides with the number S defined in the previous section. We will always write S instead of S_1 in order to be consistent with the results of Sect. 3.4.1.

The following lemma establishes the behavior of minimizers when b changes.

Lemma 3.4.14 *For every $b > 0$,*

1. $S_b = b^{\frac{N-2}{N}} S$.
2. *If u_1 is the minimizer for S found in Theorem 3.4.1, and $v_b(x) = b^{1/2^*} u_1(x)$, the function v_b is a minimizer for S_b, namely*

$$\int_{\mathbb{R}^N} |v_b|^{2^*} \, dx = b \quad and \quad \|v_b\|^2 = S_b.$$

3. *There results*

$$S_b = \inf \left\{ \|u\|^2 \;\middle|\; u \in D_r, \int_{\mathbb{R}^N} |u|^{2^*} \, dx = b \right\}.$$

Proof As it has been sketched in Remark 2.3.6, it is easy to see that for all $a, b > 0$ there results

$$\frac{S_a}{a^{\frac{N-2}{N}}} = \frac{S_b}{b^{\frac{N-2}{N}}},$$

so for $a = 1$ we obtain the claim. The proof of the second part is obvious, and the third part easily derives from the fact that u_1 and v_b are radial functions. $\square$

We now define the number

$$\overline{S} = \inf\left\{ \|u\|^2 \;\middle|\; u \in D_r, \int_{\mathbb{R}^N} \beta(|x|)|u|^{2^*} dx = 1 \right\}$$

and we show that $\overline{S}$ is attained.

We start the argument by setting, in case $\beta(0) > 0$,

$$\beta_0 = \beta(0)$$

and we first compare $\overline{S}$ to $S_{\beta_0^{-1}}$.

Lemma 3.4.15 *We have $\overline{S} < S_{\beta_0^{-1}}$.*

Proof Let w be a nonnegative radial minimizer for $S_{\beta_0^{-1}}$, that is, $w \in D_r \setminus \{0\}$, $w \geq 0$,

$$\int_{\mathbb{R}^N} \beta_0 w^{2^*} dx = 1 \quad \text{and} \quad \|w\|^2 = S_{\beta_0^{-1}}.$$

From (**b₂**) we see that

$$\int_{\mathbb{R}^N} \beta(|x|)w^{2^*} dx > \int_{\mathbb{R}^N} \beta_0 w^{2^*} dx = 1.$$

We define now

$$\mu = \left(\int_{\mathbb{R}^N} \beta(|x|)w^{2^*} dx \right)^{1/2^*},$$

so that $\mu > 1$, and $v = w/\mu$. Then

$$\int_{\mathbb{R}^N} \beta(|x|)v^{2^*} dx = 1$$

and therefore

$$\overline{S} \leq \|v\|^2 = \frac{1}{\mu^2}\|w\|^2 < S_{\beta_0^{-1}}. \qquad \square$$

We can now repeat the arguments of the previous section, and we give the details only in case something different has to be done. As a first thing, arguing as we did in Lemmas 3.4.5 and 3.4.6, we obtain the following result.

Lemma 3.4.16 *There exist a minimizing sequence $\{u_k\}_k \subset D_r$ for $\overline{S}$ and a function $u \in D_r$ such that*

- $u_k \geq 0$, $u \geq 0$.
- $u_k \rightharpoonup u$ in $D^{1,2}(\mathbb{R}^N)$ and in $L^{2^*}(\mathbb{R}^N)$, and $u_k(x) \to u(x)$ a.e. in $\mathbb{R}^N$.
- for all $j \in \mathbb{N}$ the sequences

$$\left\{ \int_{|x|<1/j} |\nabla u_k|^2 \, dx \right\}_k \quad and \quad \left\{ \int_{|x|>j} |\nabla u_k|^2 \, dx \right\}_k$$

admit finite limits as $k \to \infty$.

We continue with the analogue of Lemma 3.4.7.

Lemma 3.4.17 *The following limits exist and are finite*:

1. $\nu_0 = \lim_{R \to 0^+} (\limsup_k \int_{|x|<R} \beta(|x|)|u_k|^{2^*} \, dx)$,
   $\nu_\infty = \lim_{R \to +\infty} (\limsup_k \int_{|x|\geq R} \beta(|x|)|u_k|^{2^*} \, dx)$,
2. $\mu_0 = \lim_{R \to 0^+} (\limsup_k \int_{|x|<R} |\nabla u_k|^2 \, dx)$,
   $\mu_\infty = \lim_{R \to +\infty} (\limsup_k \int_{|x|\geq R} |\nabla u_k|^2 \, dx)$.

   Moreover, $\nu_0, \nu_\infty \in [0, 1]$ *and*

3. *if* $\beta(0) = 0$ *then* $\nu_0 = \nu_\infty = 0$.

Proof The argument is exactly the same as that of Lemma 3.4.7. We have only to prove the last claim. Assume $\beta(0) = 0$ and take $\varepsilon > 0$ and $r_\varepsilon, R_\varepsilon > 0$ such that

$$\beta(r) \leq \varepsilon \quad \forall r \in [0, r_\varepsilon] \cup [R_\varepsilon, +\infty)$$

which is possible by **(b₂)**. Define

$$M = \sup_k \left\{ \int_{\mathbb{R}^N} |u_k|^{2^*} \, dx \right\}.$$

Then, for all $0 < R_1 < r_\varepsilon$ and all $R_2 > R_\varepsilon$ we have

$$\int_{|x|<R_1} \beta(|x|)|u_k|^{2^*} \, dx \leq M\varepsilon \quad and \quad \int_{|x|>R_2} \beta(|x|)|u_k|^{2^*} \, dx \leq M\varepsilon,$$

so that $\nu_0 \leq M\varepsilon$ and $\nu_\infty \leq M\varepsilon$. This holds for all $\varepsilon > 0$ and the claim follows.   □

This allows us to prove the same limit relations on the above quantities as we did in the previous section. We collect all the results in a single statement.

Lemma 3.4.18 *There results*

$$1 \leq \int_{\mathbb{R}^N} \beta(|x|)|u|^{2^*} \, dx + \nu_0 + \nu_\infty;$$

$$\overline{S} \geq \int_{\mathbb{R}^N} |\nabla u|^2 \, dx + \mu_0 + \mu_\infty;$$

$$\overline{S} \nu_0^{2/2^*} \leq \mu_0, \qquad \overline{S} \nu_\infty^{2/2^*} \leq \mu_\infty.$$

Lemma 3.4.19 *Only the following three cases can occur: either*

- $\int_{\mathbb{R}^N} \beta(|x|) u^{2^*} \, dx = 1$ and $v_0 = v_\infty = 0$,
- or $v_0 = 1$ and $\int_{\mathbb{R}^N} \beta(|x|) u^{2^*} \, dx = v_\infty = 0$,
- or $v_\infty = 1$ and $\int_{\mathbb{R}^N} \beta(|x|) u^{2^*} \, dx = v_0 = 0$.

Proof Same as that of Lemma 3.4.11.   □

The only result that requires a new proof is the next one.

Lemma 3.4.20 *There results*

$$v_0 = v_\infty = 0 \quad and \quad \int_{\mathbb{R}^N} \beta(|x|)|u|^{2^*} \, dx = 1.$$

Proof We argue by contradiction. If the claim is not true, then by the previous lemma we have

$$\int_{\mathbb{R}^N} \beta(|x|)|u|^{2^*} \, dx = 0,$$

that is $\beta(|x|)|u|^{2^*} = 0$ a.e., and $v_0 = 1$ or $v_\infty = 1$. Hence, by Lemma 3.4.17, we have $\beta(0) = \beta_0 > 0$, so, by Lemma 3.4.15, $\overline{S} < S_{\beta_0^{-1}}$. We claim that

$$\lim_k \int_{\mathbb{R}^N} \beta_0 |u_k|^{2^*} \, dx = 1. \tag{3.23}$$

Indeed we have

$$1 = \int_{\mathbb{R}^N} \beta(|x|)|u_k|^{2^*} \, dx = \int_{\mathbb{R}^N} \beta_0 |u_k|^{2^*} \, dx + \int_{\mathbb{R}^N} (\beta(|x|) - \beta_0)|u_k|^{2^*} \, dx,$$

so we have to prove that

$$\lim_k \int_{\mathbb{R}^N} (\beta(|x|) - \beta_0)|u_k|^{2^*} \, dx = 0.$$

To this aim, fix $\varepsilon > 0$ and take $r_\varepsilon, R_\varepsilon > 0$ such that $\beta(r) - \beta_0 \le \varepsilon$ for all $r \in [0, r_\varepsilon]$ and all $r \in [R_\varepsilon, +\infty)$, and define

$$M = \sup_k \left\{ \int_{\mathbb{R}^N} |u_k|^{2^*} \, dx \right\}.$$

Then

$$0 \le \int_{\mathbb{R}^N} (\beta(|x|) - \beta_0)|u_k|^{2^*} \, dx$$

$$= \int_{|x|<r_\varepsilon} (\beta(|x|) - \beta_0)|u_k|^{2^*} \, dx + \int_{r_\varepsilon<|x|<R_\varepsilon} (\beta(|x|) - \beta_0)|u_k|^{2^*} \, dx$$

$$+ \int_{|x|>R_\varepsilon} (\beta(|x|) - \beta_0)|u_k|^{2^*} \, dx$$

$$\le 2M\varepsilon + \int_{r_\varepsilon<|x|<R_\varepsilon} (\beta(|x|) - \beta_0)|u_k|^{2^*} \, dx.$$

In the bounded set $A_\varepsilon = \{x \in \mathbb{R}^N \mid r_\varepsilon < |x| < R_\varepsilon\}$ the functions $\beta(|x|)|u_k|^{2^*}$ are uniformly bounded and they converge pointwise to $\beta(|x|)|u|^{2^*} = 0$, so that, by dominated convergence,

$$\lim_k \int_{r_\varepsilon < |x| < R_\varepsilon} (\beta(|x|) - \beta_0)|u_k|^{2^*} \, dx = 0,$$

which yields

$$0 \leq \limsup_k \int_{\mathbb{R}^N} (\beta(|x|) - \beta_0)|u_k|^{2^*} \, dx \leq 2M\varepsilon.$$

Since this holds for all $\varepsilon > 0$,

$$\lim_k \int_{\mathbb{R}^N} (\beta(|x|) - \beta_0)|u_k|^{2^*} \, dx = 0,$$

which concludes the proof of (3.23). Now we can finish the proof of the lemma.
    Define $\mu_k = (\int_{\mathbb{R}^N} \beta_0 |u_k|^{2^*} \, dx)^{1/2^*}$ and $v_k = u_k/\mu_k$. We have

$$\int_{\mathbb{R}^N} \beta_0 |v_k|^{2^*} \, dx = 1$$

and therefore

$$S_{\beta_0^{-1}} \leq \|v_k\|^2 = \frac{1}{\mu_k^2}\|u_k\|^2 \to \overline{S}.$$

This shows that $S_{\beta_0^{-1}} \leq \overline{S}$, a contradiction.   □

    Since

$$\int_{\mathbb{R}^N} \beta(|x|)|u|^{2^*} \, dx = 1,$$

arguing as in Chap. 2 we can conclude that there exists $u \in D_r \setminus \{0\}$, $u \geq 0$, such that

$$\int_{\mathbb{R}^N} \nabla u \cdot \nabla v \, dx = \int_{\mathbb{R}^N} \beta(|x|)u^{2^*-1}v \, dx \tag{3.24}$$

for all $v \in D_r$. We now want to prove that this equation is satisfied for all v in $D^{1,2}(\mathbb{R}^N)$. This is carried out in the following lemma, which is a particular case of the "principle of symmetric criticality" by Palais [38], see also [48].

Lemma 3.4.21 *Equation* (3.24) *holds for every* $v \in D^{1,2}(\mathbb{R}^N)$.

Proof Let us consider the functional $I : D^{1,2}(\mathbb{R}^N) \to \mathbb{R}$ defined by

$$I(u) = \frac{1}{2}\|u\|^2 - \frac{1}{2^*}\int_{\mathbb{R}^N} \beta(|x|)|u|^{2^*} \, dx$$

and let I_r be its restriction to D_r. By (3.24) we have $I_r'(u) = 0$. We want to prove that also $I'(u) = 0$, that is, $\nabla I(u) = 0$ in $D^{1,2}(\mathbb{R}^N)$. For this, we claim that $\nabla I(u)$ is a *radial* function.

To prove the claim, we take a rotation of $\mathbb{R}^N$ with matrix R and we denote its transpose by S. Of course, $S = R^{-1}$.

We set $f_R(x) = f(Rx)$ for all $f \in D^{1,2}(\mathbb{R}^N)$ and we prove that for all such R and for all $f, g \in D^{1,2}(\mathbb{R}^N)$ there results

$$\int_{\mathbb{R}^N} \nabla f_R \cdot \nabla g \, dx = \int_{\mathbb{R}^N} \nabla f \cdot \nabla g_S \, dx. \tag{3.25}$$

We check (3.25) for smooth f, g, the general case being easily obtained by density. An easy computation gives

$$\nabla f_R(x) = S \nabla f(Rx),$$

and hence

$$\nabla f_R(x) \cdot \nabla g(x) = S \nabla f(Rx) \cdot \nabla g(x) = \nabla f(Rx) \cdot R \nabla g(x).$$

Integrating and changing variables yields

$$\int_{\mathbb{R}^N} \nabla f_R(x) \cdot \nabla g(x) \, dx = \int_{\mathbb{R}^N} \nabla f(Rx) \cdot R \nabla g(x) \, dx = \int_{\mathbb{R}^N} \nabla f(x) \cdot R \nabla g(Sx) \, dx$$

$$= \int_{\mathbb{R}^N} \nabla f(x) \cdot \nabla g_S(x) \, dx.$$

This means

$$(f_R | g) = (f | g_S),$$

where $(\cdot | \cdot)$ is the scalar product in $D^{1,2}(\mathbb{R}^N)$.

To prove that $\nabla I(u)$ is radial we have to prove that, for all rotations R,

$$\nabla I(u) = (\nabla I(u))_R,$$

that is

$$(\nabla I(u) | v) = \big((\nabla I(u))_R \,|\, v\big)$$

for all $v \in D^{1,2}(\mathbb{R}^N)$. Recalling that u is a radial function, we compute

$$\big((\nabla I(u))_R \,|\, v\big) = (\nabla I(u) | v_S) = I'(u) v_S$$

$$= (u | v_S) - \int_{\mathbb{R}^N} \beta(|x|) u^{2^*-1} v_S \, dx = (u_R | v) - \int_{\mathbb{R}^N} \beta(|x|) u_R^{2^*-1} v \, dx$$

$$= (u | v) - \int_{\mathbb{R}^N} \beta(|x|) u^{2^*-1} v \, dx = I'(u) v = (\nabla I(u) | v).$$

This shows that $\nabla I(u)$ is radial.

Now it is easy to conclude the proof. We have

$$0 = I'_r(u)(\nabla I(u)) = I'(u)(\nabla I(u)) = (\nabla I(u) \,|\, \nabla I(u)) = \|\nabla I(u)\|^2,$$

so $\nabla I(u) = 0$, namely $I'(u) v = 0$ for all $v \in D^{1,2}(\mathbb{R}^N)$. $\qquad\square$

Remark 3.4.22 Hypothesis (**b₂**) can be weakened in the sense that it is not necessary to assume $\beta(r) > \beta(0)$ for all $r > 0$ but it is enough that $\beta(r) \geq \beta(0)$ for all $r > 0$ and $\beta(r_0) > \beta(0)$ for some $r_0 > 0$. Indeed, if we assume this weaker condition, as we know that the minimizer for S is everywhere strictly positive (see Remark 3.4.12), in Lemma 3.4.15 we obtain

$$\int_{\mathbb{R}^N} \beta(|x|) w^{2^*} \, dx > \int_{\mathbb{R}^N} \beta_0 w^{2^*} \, dx,$$

and this is all we need to conclude.

3.4.3 A Nonexistence Result

If one tries to attack the analogue of problem (3.13) on a *bounded* domain Ω, namely the problem

$$\begin{cases} -\Delta u = |u|^{2^*-2}u & \text{in } \Omega, \\ u = 0 & \text{on } \partial\Omega, \end{cases} \tag{3.26}$$

one realizes quickly that the methods used in Sect. 3.4.1 break down.

The question that one has to face is then: are there more powerful techniques that allow one to solve (3.26), or does something really happen when the growth of the nonlinearity reaches the critical value 2*?

In this section we show, via a classical argument, that no method can produce a nontrivial solution to (3.26), for example when Ω is a ball. In other words, when power nonlinearities are considered, the value 2* represents a threshold beyond which existence of nontrivial solutions may fail. For the sake of simplicity we limit ourselves to the nonexistence of solutions *of constant sign*; the general case requires some technical arguments that are beyond the scope of this book.

We begin by describing the class of sets Ω to which the argument applies.

Definition 3.4.23 We say that an open subset Ω of $\mathbb{R}^N$ is starshaped with respect to a point $x_0 \in \Omega$ if for every $x \in \overline{\Omega}$, the segment joining x_0 to x, namely the set

$$\{\lambda x + (1 - \lambda)x_0 \mid \lambda \in [0, 1]\}$$

is entirely contained in $\overline{\Omega}$.

One says that Ω is strictly starshaped with respect to $x_0 \in \Omega$ if for every $x \in \overline{\Omega}$, the segment

$$\{\lambda x + (1 - \lambda)x_0 \mid \lambda \in [0, 1)\}$$

is entirely contained in Ω.

Normally one assumes that $0 \in \Omega$, and requires Ω to be strictly starshaped with respect to 0. In this case the definition reads

$$x \in \overline{\Omega} \quad \text{implies} \quad \lambda x \in \Omega \quad \forall \lambda \in [0, 1).$$

Convex sets are starshaped with respect to every internal point, and strict convexity implies strict starshapedness. However it is easy to see that there are starshaped sets that are not convex (those that look like a star, exactly).

The main property that we are going to use is contained in the following lemma. The reader can prove it as an exercise or consult [18].

Lemma 3.4.24 *Assume that $\Omega \subset \mathbb{R}^N$ is smooth and strictly starshaped with respect to 0. Let $v(x)$ denote the outward normal to $\partial\Omega$ at x. Then*

$$v(x) \cdot x > 0 \quad \forall x \in \partial\Omega.$$

The main nonexistence result is the following.

Theorem 3.4.25 *Assume that $\Omega \subset \mathbb{R}^N$, with $N \geq 3$, is open, bounded, smooth and strictly starshaped with respect to 0. Then the problem*

$$\begin{cases} -\Delta u = |u|^{2^*-2}u & \text{in } \Omega, \\ u > 0 & \text{in } \Omega, \\ u = 0 & \text{on } \partial\Omega \end{cases} \tag{3.27}$$

has no solutions in $H_0^1(\Omega)$.

The proof of this theorem is based on a regularity result, which asserts that any weak $H_0^1(\Omega)$ solution is in fact of class C^2 on $\overline{\Omega}$ (see [45]), and on an integral identity established in [39] that we now describe. If one wants to bypass the regularity argument, one can read the above theorem simply as follows: problem (3.27) has no smooth solutions. Of course the same result holds for negative solutions, with the same proof.

Theorem 3.4.26 (Pohožaev identity [39]) *Let $f : \mathbb{R} \to \mathbb{R}$ be continuous and let $F(t) = \int_0^t f(s)\,ds$. Assume that $\Omega \subset \mathbb{R}^N$ is open and bounded. If $u \in C^2(\overline{\Omega})$ is a solution of the problem*

$$\begin{cases} -\Delta u = f(u) & \text{in } \Omega, \\ u = 0 & \text{on } \partial\Omega, \end{cases} \tag{3.28}$$

then

$$\frac{N-2}{2}\int_\Omega |\nabla u|^2\,dx - N\int_\Omega F(u)\,dx = -\int_{\partial\Omega}\left(\frac{\partial u}{\partial v}\right)^2 v(x) \cdot x\,d\sigma. \tag{3.29}$$

Proof The proof consists in multiplying the equation in (3.28) by $\nabla u \cdot x$ and integrating by parts by means of the Green's formula (Theorem 1.2.5). We begin with the left-hand side. We have

$$-\int_\Omega \Delta u\,\nabla u \cdot x\,dx = \int_\Omega \nabla u \cdot \nabla(\nabla u \cdot x)\,dx - \int_{\partial\Omega}\frac{\partial u}{\partial v}\nabla u(x) \cdot x\,d\sigma. \tag{3.30}$$

Now

$$\frac{\partial}{\partial x_j}(\nabla u \cdot x) = \frac{\partial}{\partial x_j}\left(\sum_{i=1}^{N}\frac{\partial u}{\partial x_i}x_i\right) = \sum_{i=1}^{N}\left(\frac{\partial^2 u}{\partial x_j \partial x_i}x_i + \frac{\partial u}{\partial x_i}\delta_{ij}\right)$$

$$= \sum_{i=1}^{N}\frac{\partial^2 u}{\partial x_j \partial x_i}x_i + \frac{\partial u}{\partial x_j},$$

so that

$$\nabla u \cdot \nabla(\nabla u \cdot x) = \sum_{j=1}^{N}\frac{\partial u}{\partial x_j}\left(\sum_{i=1}^{N}\frac{\partial^2 u}{\partial x_j \partial x_i}x_i + \frac{\partial u}{\partial x_j}\right)$$

$$= \sum_{i=1}^{N}\frac{\partial}{\partial x_i}\left(\frac{1}{2}\sum_{j=1}^{N}\left(\frac{\partial u}{\partial x_j}\right)^2\right)x_i + |\nabla u|^2$$

$$= \frac{1}{2}\nabla(|\nabla u|^2)\cdot x + |\nabla u|^2 = \frac{1}{2}\nabla(|\nabla u|^2)\cdot\nabla\left(\frac{1}{2}|x|^2\right) + |\nabla u|^2.$$

Using again the Green's formula we obtain

$$\int_\Omega \nabla u \cdot \nabla(\nabla u \cdot x)\,dx$$

$$= \int_\Omega |\nabla u|^2\,dx + \frac{1}{2}\int_\Omega \nabla(|\nabla u|^2)\cdot\nabla\left(\frac{1}{2}|x|^2\right)dx$$

$$= \int_\Omega |\nabla u|^2\,dx + \frac{1}{2}\int_{\partial\Omega}|\nabla u|^2\frac{\partial}{\partial\nu}\left(\frac{1}{2}|x|^2\right)d\sigma - \frac{N}{2}\int_\Omega |\nabla u|^2\,dx$$

$$= \frac{2-N}{2}\int_\Omega |\nabla u|^2\,dx + \frac{1}{2}\int_{\partial\Omega}|\nabla u|^2\nu(x)\cdot x\,d\sigma, \qquad (3.31)$$

because $\Delta(\frac{1}{2}|x|^2) = N$.

Now we notice that since $u = 0$ on $\partial\Omega$, we have $\nabla u(x) = \frac{\partial u}{\partial\nu}(x)\nu(x)$ for every $x \in \partial\Omega$, so that $|\nabla u| = |\frac{\partial u}{\partial\nu}|$ and $\nabla u \cdot x = \frac{\partial u}{\partial\nu}\nu(x)\cdot x$ on $\partial\Omega$. Taking this into account and inserting (3.31) into (3.30) we see that

$$-\int_\Omega \Delta u\,\nabla u\cdot x\,dx = \frac{2-N}{2}\int_\Omega |\nabla u|^2\,dx + \frac{1}{2}\int_{\partial\Omega}|\nabla u|^2\nu(x)\cdot x\,d\sigma$$

$$- \int_{\partial\Omega}\frac{\partial u}{\partial\nu}\nabla u(x)\cdot x\,d\sigma$$

$$= \frac{2-N}{2}\int_\Omega |\nabla u|^2\,dx - \frac{1}{2}\int_{\partial\Omega}\left(\frac{\partial u}{\partial\nu}\right)^2\nu(x)\cdot x\,d\sigma. \qquad (3.32)$$

The treatment of the right-hand side is much easier. Indeed we have

$$f(u)\nabla u\cdot x = \nabla(F(u))\cdot x = \nabla(F(u))\cdot\nabla\left(\frac{1}{2}|x|^2\right),$$

so that by the Green's formula,

$$\int_\Omega f(u)\nabla u \cdot x\, dx = \int_\Omega \nabla(F(u)) \cdot \nabla\left(\frac{1}{2}|x|^2\right) dx$$

$$= \int_{\partial\Omega} F(u)\frac{\partial}{\partial\nu}\left(\frac{1}{2}|x|^2\right) d\sigma - N\int_\Omega F(u)\, dx$$

$$= -N\int_\Omega F(u)\, dx, \tag{3.33}$$

since $F(u(x)) = F(0) = 0$ on $\partial\Omega$. Putting together (3.32) and (3.33), we conclude that multiplying the equation by $\nabla u \cdot x$ yields

$$\frac{N-2}{2}\int_\Omega |\nabla u|^2 dx - N\int_\Omega F(u)\, dx = -\int_{\partial\Omega}\left(\frac{\partial u}{\partial\nu}\right)^2 v(x)\cdot x\, d\sigma,$$

as we wanted to prove.   □

Proof of Theorem 3.4.25 Let $u \in H_0^1(\Omega)$ solve (3.27). By regularity results (see [45]), we can assert that $u \in C^2(\overline{\Omega})$. Applying Theorem 3.4.26, with

$$f(u) = |u|^{2^*-2}u \quad \text{and} \quad F(u) = \frac{1}{2^*}|u|^{2^*}$$

yields

$$\frac{N-2}{2}\int_\Omega |\nabla u|^2 dx - \frac{N}{2^*}\int_\Omega |u|^{2^*} dx = -\int_{\partial\Omega}\left(\frac{\partial u}{\partial\nu}\right)^2 v(x)\cdot x\, d\sigma.$$

Multiplying the equation in (3.27) by u and integrating shows that

$$\int_\Omega |\nabla u|^2 dx = \int_\Omega |u|^{2^*} dx,$$

which we can insert in the preceding formula to obtain

$$\left(\frac{N-2}{2} - \frac{N}{2^*}\right)\int_\Omega |u|^{2^*} dx = -\int_{\partial\Omega}\left(\frac{\partial u}{\partial\nu}\right)^2 v(x)\cdot x\, d\sigma.$$

But

$$\frac{N-2}{2} - \frac{N}{2^*} = 0$$

by definition of 2^*, so that

$$\int_{\partial\Omega}\left(\frac{\partial u}{\partial\nu}\right)^2 v(x)\cdot x\, d\sigma = 0.$$

Since Ω is strictly starshaped with respect to 0, we have $v(x)\cdot x > 0$ everywhere on $\partial\Omega$, and then, necessarily,

$$\frac{\partial u}{\partial\nu}(x) = 0 \quad \forall x \in \partial\Omega.$$

We conclude by integrating the equation in (3.27) and using again the Green's formula. We obtain

$$\int_\Omega |u|^{2^*-2}u\, dx = -\int_\Omega \Delta u\, dx = -\int_{\partial\Omega}\frac{\partial u}{\partial\nu}\, d\sigma = 0.$$

Since u is positive in Ω, this is impossible.   □

Remark 3.4.27 What about non starshaped domains? This question has been the object of intense research starting in the eighties. Celebrated results (see the references at the end of the chapter) first showed that if the set Ω is an annulus, or has a small hole, then problem (3.27) has a solution. The most general result states that if Ω has nontrivial topology in some appropriate sense, then the problem is still solvable. So for a while it seemed that the relevant point was the topology of Ω. But then other results showed that the problem still has a solution when Ω has trivial topology (e.g., is contractible), but *looks like* a noncontractible domain, for example an annulus with a thin radial slice removed. Summing up, the solvability of problems with critical growth is a *very* delicate matter and depends subtly on the data of the problem.

3.5 Exercises

1. Assume that $N \geq 3$. Prove that there exists a constant $C > 0$ such that

    $$\int_{\mathbb{R}^N} \frac{u^2}{|x|^2} \, dx \leq C \int_{\mathbb{R}^N} |\nabla u|^2 \, dx \quad \forall u \in D^{1,2}(\mathbb{R}^N). \tag{3.34}$$

    This is the celebrated *Hardy inequality*. *Hint*: assume first that $u \in C_0^\infty(\mathbb{R}^N)$ and notice that left-hand side is finite. Write, for $\lambda > 0$ and $x \in \mathbb{R}^N$,

    $$u^2(x) = -\int_1^{+\infty} \frac{d}{d\lambda} u^2(\lambda x) \, d\lambda = -2 \int_1^{+\infty} u(\lambda x) \nabla u(\lambda x) \cdot x \, d\lambda.$$

    Then divide by $|x|^2$ and integrate over $\mathbb{R}^N$ setting $y = \lambda x$. Conclude by density.
2. Use the Hardy inequality (3.34) to prove that the expression

    $$(u|v) = \int_{\mathbb{R}^N} \nabla u \cdot \nabla v \, dx + \int_{\mathbb{R}^N} \frac{1}{|x|^2} uv \, dx$$

    defines a scalar product in $D^{1,2}$ that induces a norm equivalent to the standard one. Then
    (a) Let $\|u\| = (u|u)^{1/2}$. For every $u \in D^{1,2}(\mathbb{R}^N)$ and every $\lambda > 0$ define

    $$u_\lambda(x) = \lambda^{\frac{N-2}{2}} u(\lambda x).$$

    Prove that $\|u_\lambda\| = \|u\|$ for every $\lambda > 0$.
    (b) Define

    $$C = \inf \left\{ \|u\|^2 \mid u \in D^{1,2}(\mathbb{R}^N), \int_\Omega |u|^{2^*} = 1 \right\}.$$

    Prove that $C > 0$.
3. Using the notation and the results of Exercise 2, find a radial non negative and non trivial solution of the problem

    $$\begin{cases} -\Delta u + \dfrac{1}{|x|^2} u = |u|^{2^*-2} u, \\ \\ u \in D^{1,2}(\mathbb{R}^N). \end{cases}$$

Hint: define $D_r = \{u \in D^{1,2}(\mathbb{R}^N) \mid u \text{ is radial}\}$,

$$\mathcal{R} = \inf\left\{ \|u\|^2 \,\Big|\, u \in D_r, \int_\Omega |u|^{2^*} = 1 \right\},$$

and adapt the lemmas and arguments of the proof of Theorem 3.4.1. In the paper [47] the reader can find the complete classification of radial and positive solutions of this problem.

4. Let $f : \mathbb{R} \to \mathbb{R}$ be a continuous function such that

$$|f(t)| \le C\, |t|^{2^*-1} \quad \forall t \in \mathbb{R}$$

for some $C > 0$. Define $F(t) = \int_0^t f(s)\,ds$ and $J : D^{1,2}(\mathbb{R}^N) \to \mathbb{R}$ as

$$J(u) = \int_{\mathbb{R}^N} F(u)\,dx.$$

   (a) Prove that J is a differentiable functional with

$$J'(u)v = \int_{\mathbb{R}^N} f(u)v\,dx \quad \forall u, v \in D^{1,2}(\mathbb{R}^N).$$

   (b) Define $M = \{u \in D^{1,2}(\mathbb{R}^N) \mid \int_{\mathbb{R}^N} F(u)\,dx = 1\}$, assume $M \ne \emptyset$ and define $m = \inf\{\|u\|^2 \mid u \in M\}$, where $\|u\|$ is the norm in $D^{1,2}(\mathbb{R}^N)$. Prove that $m > 0$.

   (c) Assume that there exists $u \in M$ such that

$$\|u\|^2 = m \quad \text{and} \quad \int_{\mathbb{R}^N} f(u)u\,dx \ne 0.$$

   Prove that u is a weak solution of the nonlinear eigenvalue problem

$$\begin{cases} -\Delta v = \lambda f(v), \\ v \in D^{1,2}(\mathbb{R}^N). \end{cases}$$

   Hint: define $\varphi(s,t) = \int_{\mathbb{R}^N} F(t(u + sv))dx$, for (s,t) in a neighborhood of $(0,1)$.

   (d) For every $\sigma > 0$ define $u_\sigma(x) = u(x/\sigma)$. Prove that if $\int_{\mathbb{R}^N} f(u)u\,dx > 0$, then there is a $\sigma > 0$ such that u_σ is a solution of the problem

$$\begin{cases} -\Delta v = f(v), \\ v \in D^{1,2}(\mathbb{R}^N). \end{cases}$$

5. Let $2 < p < 2^* < q$ and define a continuous function $f : \mathbb{R} \to \mathbb{R}$ by

$$f(t) = \min\{t^{p-1}, t^{q-1}\} \quad \text{for } t \ge 0, \qquad f(-t) = -f(t) \quad \text{for } t < 0.$$

Define $F(t) = \int_0^t f(s)\,ds$.

   (a) Prove that $|f(t)| \le |t|^{2^*-1}$ for all $t \in \mathbb{R}$, and compute $F(t)$.

   (b) Consider a sequence $\{u_k\} \subset D_r$ such that $u_k \rightharpoonup u$ in D_r, and prove that

$$\int_{\mathbb{R}^N} F(u_k)\,dx \to \int_{\mathbb{R}^N} F(u)\,dx.$$

Hint: notice that both u_k and u satisfy the inequality $|v(x)| \leq C/|x|^{\frac{N-2}{2}}$, for a suitable $C > 0$. Hence you can fix $R > 0$ such that both u_k and u satisfy $|v(x)| < 1$ if $|x| > R$. Write

$$\int_{\mathbb{R}^N} F(u_k)\,dx = \int_{|x| \leq R} F(u_k)\,dx + \int_{|x| > R} F(u_k)\,dx$$

and apply dominated convergence in the right-hand side integrals, noticing also that $|F(t)| \leq a + b|t|^p$ for suitable $a, b > 0$ and all t.

(c) Define $M_r = \{u \in D_r \mid \int_{\mathbb{R}^N} F(u)\,dx = 1\}$. Prove that $M_r \neq \emptyset$.

(d) Define

$$m_r = \inf \{\|u\|^2 \mid u \in M_r\},$$

where $\|u\|$ is the norm in $D^{1,2}(\mathbb{R}^N)$. Notice that, as in the previous exercise, $m_r > 0$. Show that there exists $u \in M_r$, $u \geq 0$, such that $\|u\|^2 = m_r$.

(e) Using the arguments of the previous exercise and of Sect. 3.4.2, prove that there exists a non trivial and non negative solution of the problem

$$\begin{cases} -\Delta v = f(v) \\ v \in D_r. \end{cases}$$

6. Let $q : \mathbb{R}^N \to \mathbb{R}$ be continuous, strictly positive and such that $\lim_{|x| \to \infty} q(x) = 0$. For $\lambda \geq 0$ and $p \in (2, 2^*)$, consider the problems

$$\begin{cases} -\Delta u + (1 + \lambda q(x))u = |u|^{p-2}u, \\ u \in H^1(\mathbb{R}^N). \end{cases}$$

Weak solutions of these problems are, for example, critical points of the functionals $Q_\lambda : H^1(\mathbb{R}^N) \setminus \{0\} \to \mathbb{R}$ defined by

$$Q_\lambda(u) = \frac{\int_{\mathbb{R}^N} |\nabla u|^2\,dx + \int_{\mathbb{R}^N}(1 + \lambda q(x))u^2\,dx}{(\int_{\mathbb{R}^N} |u|^p\,dx)^{2/p}}.$$

(a) Prove that for every $\lambda > 0$,

$$\inf_{u \in H^1(\mathbb{R}^N) \setminus \{0\}} Q_\lambda(u) = \inf_{u \in H^1(\mathbb{R}^N) \setminus \{0\}} Q_0(u).$$

(b) Prove that for every $\lambda > 0$, Q_λ has no global minimum on $H^1(\mathbb{R}^N) \setminus \{0\}$.

7. Let $p \in (2, 2^*)$ and let $a : \mathbb{R}^N \to \mathbb{R}$ be the function $a(x) = \frac{1+|x|^2}{2+|x|^2}$. Prove that the functional $Q : H^1(\mathbb{R}^N) \setminus \{0\} \to \mathbb{R}$ defined by

$$Q(u) = \frac{\int_{\mathbb{R}^N} |\nabla u|^2\,dx + \int_{\mathbb{R}^N} u^2\,dx}{(\int_{\mathbb{R}^N} a(x)|u|^p)^{2/p}\,dx}$$

has no global minimum.

8. Let $U : \mathbb{R}^N \to \mathbb{R}$, with $N \geq 3$, be defined by

$$U(x) = \frac{C}{(1 + |x|^2)^{\frac{N-2}{2}}},$$

where C is a positive constant. Show that $U \in D^{1,2}(\mathbb{R}^N)$ and that U solves

$$-\Delta U = U^{2^*-1} \quad \text{in } \mathbb{R}^N \tag{3.35}$$

if and only if $C = (N(N-2))^{\frac{N-2}{4}}$. Show that for every $y \in \mathbb{R}^N$ and every $\lambda > 0$, the scaled functions

$$U_{\lambda,y}(x) = \lambda^{\frac{N-2}{2}} U(\lambda(x-y))$$

are all solutions of (3.35).

9. Assume that Ω is as in Theorem 3.4.25. Show that the problem

$$\begin{cases} -\Delta u = |u|^{p-2}u & \text{in } \Omega, \\ u = 0 & \text{on } \partial\Omega \end{cases}$$

has no $C^2(\overline{\Omega})$ solutions if $p > 2^*$.

10. Assume that Ω is as in Theorem 3.4.25 and let $q \in C^1(\overline{\Omega})$. Show that the problem

$$\begin{cases} -\Delta u + q(x)u = |u|^{2^*-2}u & \text{in } \Omega, \\ u = 0 & \text{on } \partial\Omega \end{cases}$$

has no $C^2(\overline{\Omega})$ solutions if $\nabla q(x) \cdot x + 2q(x) > 0$ in Ω.

3.6 Bibliographical Notes

- Section 3.1: The problems dealt with in the first three sections of this chapter belong to the family of *nonlinear Schrödinger equations*, appearing for example in nonlinear optics. This is an intense and active field of research at the time of this writing. The reader who wishes to broaden his/her knowledge on these topics is referred to Ambrosetti and Malchiodi [3], and to the references therein.

- Section 3.3: Problems like (3.6), and more generally subcritical problems on unbounded domains are studied extensively in Lions [31, 32], where the classical tool of concentration–compactness was introduced. A book entirely devoted to problems on unbounded domains is Kuzin and Pohožaev [27]. The type of equations described in this section contains the so called scalar field equations of physics; a more complete treatment can be found in the papers by Berestycki and Lions [9, 10].

- Section 3.4: The paper that more than any other motivated research on problems with critical growth is Brezis and Nirenberg [14]. The interested reader can find many other examples and results in Brezis [12, 13]. The effect of the topology of the domain on the existence of solutions for problems with critical exponent is a deep phenomenon. Cornerstones in this sense are the work by Coron [15] (existence on domains with a small hole), and the difficult Bahri and Coron [6] (existence when the topology of the domain is nontrivial). An example of a contractible domain on which problem (3.26) admits a solution can be found in Dancer [16].

Chapter 4
Introduction to Minimax Methods

This chapter is an introduction to a broad class of methods that have been shown to be extremely useful in a variety of contexts.

We confine ourselves to the simplest cases, but we will try to motivate the ideas involved in the construction of the main tools, so that the interested reader can turn to the study of more complex problems with a minimum of background.

In the preceding sections we have (almost) always looked for critical points of a functional as minimum points, either on the whole space, or suitably restricting the functional to sets where minima could be shown to exist. Roughly speaking, minimax methods are devoted to the search of critical points that are not global minima, for example saddle points. The procedure to do this is quite elaborate, and we will introduce the main steps gradually.

We begin with the following example, which has been, historically, one of the main motivations for the development of the theory.

Suppose that Ω is a bounded open set of $\mathbb{R}^N$, with $N \geq 3$, and that we want to find a solution u of the problem

$$\begin{cases} -\Delta u + q(x)u = |u|^{p-2}u & \text{in } \Omega, \\ u = 0 & \text{on } \partial\Omega, \end{cases} \tag{4.1}$$

where q is continuous and nonnegative and $p \in (2, 2^*)$. We know that solutions arise as critical points of the functional

$$J(u) = \frac{1}{2}\int_\Omega |\nabla u|^2\,dx + \frac{1}{2}\int_\Omega q(x)u^2\,dx - \frac{1}{p}\int_\Omega |u|^p\,dx = \frac{1}{2}\|u\|^2 - \frac{1}{p}|u|_p^p.$$

Of course, we already know at least two methods that can be used to solve Problem (4.1), and these are minimization on spheres or on the Nehari manifold (both discussed in Sect. 2.3). These methods allow one to overcome the fact that $\inf_{H_0^1(\Omega)} J = -\infty$ by *constraining* J on suitable sets where it becomes bounded from below; at this point minimization arguments can be profitably used.

We now wish to take a different point of view, which consists in looking at J as a *free* (unconstrained) functional on the whole space $H_0^1(\Omega)$.

M. Badiale, E. Serra, *Semilinear Elliptic Equations for Beginners*, Universitext, DOI 10.1007/978-0-85729-227-8_4, © Springer-Verlag London Limited 2011

We start by noticing that since (4.1) admits the trivial solution $u \equiv 0$, the function $u \equiv 0$ must be a critical point of J. What kind of critical point is it? Or, in other words, what is the variational characterization of the trivial solution?

To answer these questions we recall that by the Sobolev inequalities and our assumptions on p, there exists a constant $C > 0$ such that

$$|u|_p \leq C\|u\|, \quad \text{for every } u \in H_0^1(\Omega).$$

Therefore,

$$J(u) = \frac{1}{2}\|u\|^2 - \frac{1}{p}|u|_p^p \geq \frac{1}{2}\|u\|^2 - \frac{C^p}{p}\|u\|^p \geq \|u\|^2\left(\frac{1}{2} - \frac{C^p}{p}\|u\|^{p-2}\right).$$

Since $p > 2$, this shows that if $\|u\|$ is small enough, $J(u) \geq 0 = J(0)$ (strictly if $u \neq 0$). Thus the trivial solution, seen as a critical point of J, is a *strict local minimum*.

The second remark is that since J is unbounded from below, there certainly exists a point v, as far as we like from zero, where $J(v) < 0$.

Now if J were defined on a one-dimensional space, this would imply at once the existence of at least a *second* critical point; to visualize it think of the real function $f(x) = x^2 - x^4$, with $x \in \mathbb{R}$.

On an n-dimensional space the existence of a strict local minimum for a function unbounded from below suggests the existence of a second critical point, although this is in general false. As an example where everything works fine, we can consider in the plane the function $g(x, y) = x^2 + y^2 - x^4$.

The idea is to try to carry this "guess" to an infinite dimensional context. To do this we will have to introduce a variety of concepts and to reason about the link between the topological properties of certain sets and the existence of critical points. These concepts will be introduced at a rather abstract level in the following section.

4.1 Deformations

We begin with a definition; we work for the moment in the context of Banach spaces, while later, when we need more structure, we will turn to Hilbert spaces.

Definition 4.1.1 Let X be a Banach space and let $J : X \to \mathbb{R}$ be a functional. For $a \in \mathbb{R}$, we define

$$J^a = \{u \in X \mid J(u) \leq a\}.$$

The sets J^a are called the sublevel sets of J, or simply the sublevels of J. Any point $u \in J^{-1}(a)$ is said to be a point at level a.

In the sequel we will be interested in topological properties of the sublevel sets of a functional and, mostly, in the relationships between topological properties of sublevels and presence of critical points. To make this clearer we consider an example in dimension one.

Example 4.1.2 Consider the function $f : \mathbb{R} \to \mathbb{R}$ defined by $f(x) = x^3 - 3x$. It has two critical points, $x = \pm 1$ at the levels ∓ 2, respectively. For $a > 2$ or $a < -2$, the sublevels f^a are intervals of the type $(-\infty, x_0]$ (x_0 depends on a of course). If, on the contrary, $a \in (-2, 2)$, then f^a is a set of the type $(-\infty, x_1] \cup [x_2, x_3]$, with $x_1 < x_2$. So, for $|a| > 2$ the sublevels f^a are connected, while for $|a| < 2$ they are disconnected. This is expressed by saying that there is a *change of topology* in the sublevels when a passes through ± 2, which are exactly the levels of the critical points of f.

On the other hand, it is easy to accept (visually, for smooth functions of one variable) that if there are no critical points with levels in $[a, b]$, then f^a and f^b should "have the same topology". This is because "nothing happens" to the graph of f in the strip $\mathbb{R} \times [a, b]$.

The previous example suggests that a change in the topology of the sublevel sets of a function implies the presence of a critical point. Unfortunately, this elegant principle is false.

Example 4.1.3 Draw the graph the real function $f(x) = xe^{1-x}$. For $a \in (0, 1)$ the sublevels f^a are all disconnected, while for $a < 0$ they are connected. Yet there is no critical point at level zero.

Looking carefully at this example shows that actually there is some kind of critical point at level zero also in this case. Indeed there exists a sequence x_k such that $f(x_k) \to 0$ and $f'(x_k) \to 0$. The problem is that this sequence of "almost critical points" escapes to $+\infty$, namely it is divergent. One often says that there is a critical point *at infinity*, where the precise meaning of this term is the existence of a sequence like x_k.

This type of sequences are very important in Critical Point Theory, and they deserve a name of their own.

Definition 4.1.4 Let X be a Banach space and let $J : X \to \mathbb{R}$ be a differentiable functional. A sequence $\{u_k\}_k \subseteq X$ such that

$$\{J(u_k)\}_k \quad \text{is bounded (in } \mathbb{R}) \quad \text{and}$$
$$J'(u_k) \to 0 \quad (\text{in } X') \text{ as } k \to \infty,$$

is called a Palais–Smale sequence for J.

Let $c \in \mathbb{R}$. If

$$J(u_k) \to c \quad (\text{in } \mathbb{R}) \quad \text{and}$$
$$J'(u_k) \to 0 \quad (\text{in } X') \text{ as } k \to \infty,$$

then $\{u_k\}_k$ is called a Palais–Smale sequence for J at level c. In this case c is called a Palais–Smale level for J.

Remark 4.1.5 In a Hilbert space H we can identify the differential with the gradient via the scalar product. Therefore the second property of a Palais–Smale sequence reads

$$\nabla J(u_k) \to 0 \quad \text{in } H.$$

We point out that the convergence takes place in the *strong* topology of H.

Example 4.1.6 Consider again the real function $f(x) = xe^{1-x}$. It has only one critical point, $x = 1$ and its level is $f(1) = 1$.

The sequence $x_k = 1 - 1/k$ is a Palais–Smale sequence for f at level 1. It converges to the critical point $x = 1$. The sequence $x_k = k$ is a Palais–Smale sequence for f at level zero. It diverges to $+\infty$.

The convergence of Palais–Smale sequences is of course crucial, and is obviously a property of the functional in question.

Definition 4.1.7 Let X be a Banach space and let $J : X \to \mathbb{R}$ be a differentiable functional. We say that J satisfies the Palais–Smale condition (shortly: J satisfies (PS)) if every Palais–Smale sequence for J has a converging subsequence (in X). We say that J satisfies the Palais–Smale condition at level $c \in \mathbb{R}$ (shortly: J satisfies $(PS)_c$) if every Palais–Smale sequence at level c has a converging subsequence (in X).

The function $f(x) = xe^{1-x}$ satisfies $(PS)_1$, but not $(PS)_0$. We notice that a functional J satisfies the (PS) condition at a certain level also whenever there is no Palais–Smale sequence at that level (there is nothing to check). With this agreement, $f(x) = xe^{1-x}$ satisfies $(PS)_c$ if and only if $c \neq 0$.

The proof of the following lemma is obvious.

Lemma 4.1.8 *Let X be a Banach space and let $J : X \to \mathbb{R}$ be a C^1 functional. If there exists a Palais–Smale sequence for J and J satisfies (PS), then J has a critical point. If there exists a Palais–Smale sequence for J at level c and J satisfies $(PS)_c$, then J has a critical point at level c.*

Remark 4.1.9 Although the preceding lemma is trivial, it is of a fundamental importance for what follows. Indeed it shows that the search for critical points can be split into two independent parts: the *existence* of Palais–Smale sequences, which will follow from *topological* reasons, and the *convergence* of this sequences, which is a *compactness* problem. Moreover, as we will see, the two aspects are completely unrelated.

We now concentrate on the existence of Palais–Smale sequences, by extending and making rigorous the heuristic ideas involved in Example 4.1.2.

The first thing to do is to make clear the concept of change of topology, or rather the converse, namely that two sets have the same topology. Of course this is presented in a form that is suited for our aims.

Definition 4.1.10 Let X be a Banach space and let $B \subseteq X$ be a subset. A deformation of B is a continuous function $\eta : [0, 1] \times B \to B$ such that $\eta(0, u) = u$ for all $u \in B$.

Definition 4.1.11 Let X be a Banach space and let $A \subseteq B \subseteq X$ be subsets. We say that B is deformable in A if there exists a deformation η of B such that

$$\eta(t, u) \in A \quad \text{for all } u \in A \text{ and all } t \in [0, 1], \tag{4.2}$$

$$\eta(1, u) \in A \quad \text{for all } u \in B. \tag{4.3}$$

It is very useful to visualize the variable t as time and follow the evolution of a point as time increases. At time zero, $\eta(0, \cdot)$ is the identity on B. When t increases some points start moving, and at $t = 1$ every point of B has moved to A.

In the previous definition, a most important point is (4.2), for it says that during the deformation, the points of A can move but *cannot leave A*.

Example 4.1.12 In a Banach space X, for every $r > 0$, the ball $B_{2r}(0)$ is deformable in the ball $B_r(0)$. An explicit deformation is $\eta(t, u) = (1 - t/2)u$. More generally, any ball is deformable in any ball contained in it.

Example 4.1.13 Let X be a Banach space and let v satisfy $\|v\| = 2$. Set $B = B_4(0)$, and $A = B_1(v) \cup B_1(-v)$, so that A has two connected components. Then B is not deformable in A. Indeed, since the points of A cannot leave A during the deformation, the set $\eta(1, B)$ should be disconnected, which is impossible since it is the image of the connected set B through the continuous function $\eta(1, \cdot)$.

From now on we discuss the following problem: given a functional J on a Banach space X, and two real numbers $a < b$, when is J^b deformable in J^a? Suppose that we can find a continuous function $\eta : [0, 1] \times J^b$ such that the function $t \mapsto J(\eta(t, u))$ is nonincreasing for every u. Such function automatically satisfies $\eta(t, u) \in J^a$ for all $u \in J^a$ and all t, namely it satisfies (4.2). At first sight it seems very difficult to find a function with this property. However, in Hilbert spaces there is a very simple way to do it, through the flow associated to an autonomous differential equation. In a Banach space there are some technical questions to settle, but essentially the same ideas work. In view of our applications, we limit ourselves to the case of Hilbert spaces.

Theorem 4.1.14 *Let H be a Hilbert space and let $F : H \to H$ be a locally Lipschitz continuous function. Then, for every $u \in H$, the Cauchy problem*

$$\begin{cases} \eta' = F(\eta), \\ \eta(0) = u \end{cases} \tag{4.4}$$

has a unique solution η, which is defined in an open interval I containing 0. It is customary to think of η as a function of t and u, and write $\eta(t, u)$. In this case, η is continuous in (t, u). If F is bounded, i.e. if there is C such that $\|F(u)\| \le C$

for all $u \in H$, then $I = \mathbb{R}$. The function $\eta(t, u)$ is called the flow associated to the differential equation $\eta' = F(\eta)$, and the curves $t \mapsto \eta(t, u)$ are called the flow lines.

A proof of this result can be found in [11].

Definition 4.1.15 Let H be a Hilbert space and let $J : H \to \mathbb{R}$ be a functional. We say that $J \in C^{1,1}(H)$ if $J \in C^1(H)$ and its gradient $\nabla J : H \to H$ is locally Lipschitz continuous on H.

By Theorem 4.1.14, if $J \in C^{1,1}(H)$, the problem

$$\begin{cases} \eta'(t, u) = -\nabla J(\eta(t, u)), \\ \eta(0, u) = u \end{cases} \tag{4.5}$$

has a unique solution defined in an interval I and depending continuously on (t, u). The flow η is called the *(minus) gradient flow*, or the *steepest descent flow*, since at any time t the velocity $\eta'(t)$ points in the direction of $-\nabla J(\eta(t))$, the direction along which J decreases the fastest. If $\eta(t, u)$ is the solution of (4.5), it is easy to see that the function $\gamma(t) = J(\eta(t, u))$ is strictly decreasing, unless u is a critical point of J, in which case γ is constant. Indeed,

$$\gamma'(t) = \big(\nabla J(\eta(t, u)) \mid \eta'(t)\big) = -\big(\nabla J(\eta(t, u)) \mid \nabla J(\eta(t, u))\big)$$
$$= -\|\nabla J(\eta(t, u))\|^2 < 0,$$

if u is not a critical point of J.

Remark 4.1.16 The request that J be of class $C^{1,1}$ is used to guarantee the local uniqueness of solutions to the Cauchy problem (4.5). It is the simplest technical assumption that guarantees this property, and is rather easy to check in many concrete cases. However, the reader should keep in mind that the whole theory that we are describing can be carried out by assuming only $J \in C^1(H)$. Moreover the same theory is not confined to Hilbert spaces, but can be extended quite easily to Banach spaces, and even Banach manifolds. In all these cases one replaces ∇J with a so-called *pseudo-gradient vector field* in the Cauchy problem. The use of a pseudo-gradient vector field yields both uniqueness in (4.5) and the possibility to work without the Hilbert structure. We prefer to work in $C^{1,1}$ because the construction of a pseudo-gradient vector field is quite technical and can be considered as a more advanced topic. The interest reader will find this construction in many texts, for instance [2, 26, 42, 45].

We can now state and prove the main result of this section. It is the rigorous formalization of the heuristic ideas discussed in the first part of this section.

Lemma 4.1.17 (Deformation Lemma) *Assume that H is a Hilbert space, and let $J \in C^{1,1}(H)$. Let $a, b \in \mathbb{R}$, with $a < b$. Assume that in $[a, b]$ there is no Palais–Smale level for J, that is, for all $c \in [a, b]$ there is no Palais–Smale sequence for J at level c. Then J^b is deformable in J^a.*

Proof We construct a deformation using a variant of (4.5) to have the flow lines defined for all times. We solve the problem

$$\begin{cases} \eta'(t, u) = -\dfrac{\nabla J(\eta(t, u))}{1 + \|\nabla J(\eta(t, u))\|}, \\ \eta(0, u) = u, \end{cases} \tag{4.6}$$

where η' denotes differentiation with respect to t.

If we call $F(\eta)$ the right-hand side of (4.6), we see that $F : H \to H$ is locally Lipschitz, because ∇J is so, and of course $\|F(u)\| \leq 1$ for all u. We can then apply Theorem 4.1.14 to obtain a solution $\eta(t, u)$ of problem (4.6) defined in all of $\mathbb{R}$.

By definition we have $\eta(0, u) = u$ for all u, and since

$$\frac{d}{dt} J(\eta(t, u)) = \big(\nabla J(\eta(t, u)) \, \big| \, \eta'(t, u)\big) = -\frac{\|\nabla J(\eta(t, u))\|^2}{1 + \|\nabla J(\eta(t, u))\|} \leq 0,$$

the fact that u is in some J^c implies $\eta(t, u) \in J^c$ for all $t \geq 0$.

We now want to prove that η can be used to construct a deformation of J^b in J^a. For every $u \in J^b \backslash J^a$, we define

$$T(u) = \sup\{t \geq 0 \mid \eta(t, u) \in J^b \backslash J^a\}.$$

This could be infinite for some u, in case the flow line $\eta(\cdot, u)$ never enters J^a.

Notice that, by continuity, $T(u) > 0$ for every $u \in J^b \backslash J^a$, and for every number $t \in [0, T(u))$, we have $\eta(t, u) \in J^b \backslash J^a$.

We claim that $T(u)$ is not only finite for every u, but also uniformly bounded, in the sense that

$$\sup\{T(u) \mid u \in J^b \backslash J^a\} < +\infty. \tag{4.7}$$

To see this, we first notice that

$$\text{there exists } \delta > 0 \quad \text{such that} \quad \|\nabla J(u)\| \geq \delta \quad \text{for all } u \in J^b \backslash J^a. \tag{4.8}$$

Indeed, if (4.8) were false, then there would be a sequence $\{u_k\}_k \subset J^b \backslash J^a$ such that $\|\nabla J(u_k)\| \leq 1/k$. Passing to a subsequence, we would get $J(u_k) \to c \in [a, b]$ and $\nabla J(u_k) \to 0$; so $\{u_k\}$ would be a Palais–Smale sequence at level $c \in [a, b]$, and this is impossible by assumption.

To go back to the proof of (4.7), fix $u \in J^b \backslash J^a$ and take $t \in [0, T(u))$. Then for every $s \in [0, t]$ there results $\eta(s, u) \in J^b \backslash J^a$, so that

$$\|\nabla J(\eta(s, u))\| \geq \delta \quad \text{for all } s \in [0, t].$$

Since the function $x \mapsto -\frac{x^2}{1+x}$ is decreasing for $x \geq 0$, this implies

$$-\frac{\|\nabla J(\eta(s, u))\|^2}{1 + \|\nabla J(\eta(s, u))\|} \leq -\frac{\delta^2}{1 + \delta},$$

and this holds for every $s \in [0, t]$. Hence

$$J(\eta(t, u)) - J(\eta(0, u)) \leq \int_0^t \frac{d}{ds} J(\eta(s, u)) \, ds = \int_0^t \left(-\frac{\|\nabla J(\eta)\|^2}{1 + \|\nabla J(\eta)\|}\right) ds$$

$$\leq \int_0^t \left(-\frac{\delta^2}{1+\delta}\right) ds = -t \frac{\delta^2}{1+\delta}.$$

So, for every $t \in [0, T(u))$,

$$a < J(\eta(t, u)) \leq J(\eta(0, u)) - t \frac{\delta^2}{1+\delta} = J(u) - t \frac{\delta^2}{1+\delta} \leq b - t \frac{\delta^2}{1+\delta},$$

namely,

$$t < \frac{b-a}{\delta^2}(1+\delta).$$

Setting $\mu = \frac{b-a}{\delta^2}(1+\delta)$, we have proved that for every $t \in [0, T(u))$, there results $t < \mu$, which of course implies $T(u) \leq \mu$, and this holds for all $u \in J^b \backslash J^a$. Property (4.7) is proved.

Notice also that, by continuity, $J(\eta(T(u), u)) = a$.

We now define $\tilde{\eta} : [0, 1] \times J^b \to J^b$ by

$$\tilde{\eta}(t, u) = \eta(\mu t, u).$$

From the previous discussion $\tilde{\eta}$ enjoys the following properties:

- $\tilde{\eta}(0, u) = \eta(0, u) = u$ for all $u \in J^b$.
- $\tilde{\eta}(t, u) = \eta(\mu t, u) \in J^a$ for all $u \in J^a$.
- $J(\tilde{\eta}(1, u)) = J(\eta(\mu, u)) \leq J(\eta(T(u), u)) \leq a$ for all $u \in J^b$.

As $\tilde{\eta}$ is continuous, $\tilde{\eta}$ is a deformation of J^b that deforms J^b in J^a. $\qquad \square$

In most of the applications we will need a "local" version of the preceding result. Local here means that we only look at the functional around a certain level set.

Lemma 4.1.18 *Assume that H is a Hilbert space, let $J \in C^{1,1}(H)$, and let $c \in \mathbb{R}$. Assume that there is no Palais–Smale sequence at level c. Then for every $\varepsilon > 0$ small enough, the set $J^{c+\varepsilon}$ is deformable in $J^{c-\varepsilon}$ with a deformation that fixes $J^{c-2\varepsilon}$.*

Proof We have to find a continuous $\eta : [0, 1] \times H \to H$ such that

- $\eta(0, u) = u$ for all u,
- $\eta(t, u) = u$ for all $u \in J^{c-2\varepsilon}$ and for all t,
- $\eta(t, u) \in J^{c-\varepsilon}$ for all $u \in J^{c-\varepsilon}$ and for all t,
- $\eta(1, u) \in J^{c-\varepsilon}$ for all $u \in J^{c+\varepsilon}$.

To prove all this, we start by claiming that

there exist $\varepsilon_0 > 0$ and $\delta > 0$ such that

$$\|\nabla J(u)\| \geq \delta \quad \text{for every } u \in J^{c+\varepsilon_0} \setminus J^{c-\varepsilon_0}.$$

Indeed, if this is false then we can find positive numbers $\varepsilon_k \to 0$, $\delta_k \to 0$ and a sequence $u_k \in J^{c+\varepsilon_k} \setminus J^{c-\varepsilon_k}$ such that $\|\nabla J(u)\| \leq \delta_k$. In other words, the sequence $\{u_k\}$ satisfies

$$J(u_k) \to c \quad \text{and} \quad \nabla J(u_k) \to 0 \quad \text{in } H,$$

namely it is a Palais-sequence at level c, which is impossible by assumption.

We now pick $\varepsilon \in (0, \varepsilon_0)$, and take a "cut-off" function $\psi \in C^\infty(\mathbb{R})$ such that

- $0 \leq \psi(s) \leq 1$ for all s,
- $\psi(s) = 0$ in $(-\infty, c - 2\varepsilon]$,
- $\psi(s) = 1$ in $[c - \varepsilon, c + \varepsilon]$.

We then solve the problem

$$\begin{cases} \eta'(t, u) = -\psi(J(\eta(t, u))) \dfrac{\nabla J(\eta(t, u))}{1 + \|\nabla J(\eta(t, u))\|}, \\ \eta(0, u) = u. \end{cases} \tag{4.9}$$

Once again, the right-hand side of (4.9), is a locally Lipschitz continuous function of η and is uniformly bounded on H. Then the flow η is well defined and continuous in $\mathbb{R} \times H$.

Noticing that $\psi(J(u)) = 0$ in $J^{c-2\varepsilon}$, we have that $\eta(t, u) = u$ for all $u \in J^{c-2\varepsilon}$ and all t, namely η fixes $J^{c-2\varepsilon}$.

The other properties can be deduced observing that $\psi(J(u)) = 1$ whenever $c - \varepsilon \leq J(u) \leq c + \varepsilon$, so that one can use Lemma 4.1.17 with a and b replaced by $c - \varepsilon$ and $c + \varepsilon$ respectively. □

The previous lemma shows that the following alternative holds: either there is a Palais–Smale sequence at level c, or for every $\varepsilon > 0$ small enough, the sublevel $J^{c+\varepsilon}$ is deformable in $J^{c-\varepsilon}$ (and moreover with a deformation that "fixes" $J^{c-2\varepsilon}$).

Remark 4.1.19 There are many versions of deformations lemmas in the literature, including some very precise ones, see for example [43, 45]. For our purposes, the two above lemmas are sufficient, and are simpler than the general results.

Remark 4.1.20 In many textbooks the assumptions of the previous lemma are stated in the following equivalent form: assume that there are no critical points at level c and that $(PS)_c$ holds. We prefer the statement of Lemma 4.1.18 because it highlights the central role of Palais–Smale sequences. One can then summarize the principle by saying that *the only obstruction to deformations is the presence of Palais–Smale sequences*.

4.2 The Minimax Principle

We now introduce the minimax procedure, namely the main object of this chapter. We will do it at a rather abstract level, which has the advantage of producing a very flexible tool which can be adapted to a variety of situations, even beyond the aims of this book.

We begin with some definitions.

Definition 4.2.1 Let H be a Hilbert space and let $\eta : [0, 1] \times H \to H$ be a deformation of H. A family Γ of subsets of H is said to be invariant for η if

for every $A \in \Gamma$ and every $t \in [0, 1]$,   there results $\eta(t, A) \in \Gamma$.

Definition 4.2.2 Let H be a Hilbert space and let $J : H \to \mathbb{R}$ be a functional. A family Γ of subsets of H is called a minimax class. The value

$$c = \inf_{A \in \Gamma} \sup_{u \in A} J(u)$$

is called the minimax level associated to Γ; we also say that Γ works at level c. Notice that it may happen that $c = +\infty$ or $c = -\infty$.

Finally we have

Definition 4.2.3 Let H be a Hilbert space, $J : H \to \mathbb{R}$ a functional, Γ a minimax class, and c its minimax level. We say that Γ is admissible with respect to J if

1. $c \in \mathbb{R}$,
2. for every $\varepsilon > 0$ small enough, Γ is invariant for all deformations that fix $J^{c-2\varepsilon}$.

This last definition contains a rather subtle point. Indeed the deformations for which a class Γ should be invariant depend on c, which depends on Γ.

So the deformations for which a class Γ should be invariant depend on the class itself. This seems at first rather confusing, but we will see in the applications that the difficulty disappears as soon as the number c is determined by good topological properties. The second request in Definition 4.2.3 is extremely useful, because it reduces the number of deformations for which Γ must be invariant.

For the moment we go on at an abstract level with the main result of Critical Point Theory.

Theorem 4.2.4 (Minimax principle) *Let H be a Hilbert space and let $J \in C^{1,1}(H)$. Let Γ be an admissible minimax class at level c. Then there exists a Palais–Smale sequence for J at level c. If J satisfies* $(PS)_c$*, then there exists a critical point for J at level c.*

Proof Assume that there is no Palais–Smale sequence at level c. Then we can choose $\varepsilon > 0$ so small that Lemma 4.1.18 and 2. of Definition 4.2.3 simultaneously apply. This means that $J^{c+\varepsilon}$ is deformable in $J^{c-\varepsilon}$ with a deformation η that fixes $J^{c-2\varepsilon}$ and that Γ is invariant for this η.

By definition of minimax level, there exists a set $A \in \Gamma$ such that

$$\sup_{u \in A} J(u) \leq c + \varepsilon,$$

that is, $A \subseteq J^{c+\varepsilon}$. Applying the deformation η to A we obtain

$$\eta(1, A) \subset \eta(1, J^{c+\varepsilon}) \subset J^{c-\varepsilon},$$

which implies

$$\sup_{u \in \eta(1,A)} J(u) \leq c - \varepsilon.$$

On the other hand, as Γ is invariant for η, we have that $\eta(1, A) \in \Gamma$, so that by definition of c,

$$\sup_{u \in \eta(1,A)} J(u) \geq c.$$

This contradiction proves that there must be a Palais–Smale sequence at level c. The second part of the theorem is obvious, see Lemma 4.1.8. □

We will see some classical and particular cases of this principle in the next sections. We just point out that this principle is quite general and unifies different particular approaches that appeared separately in the literature.

Even in "trivial" contexts, it yields some nontrivial consequences, as the next example shows.

Example 4.2.5 Let H be a Hilbert space and let $J \in C^{1,1}(H)$. We consider the "trivial" minimax class

$$\Gamma = \big\{ \{u\} \mid u \in H \big\},$$

namely the class of subsets of H consisting of a single point. This class is obviously invariant for every deformation η, since $\eta(t, \{u\})$ is a singleton for every t, and hence it still belongs to Γ. Thus the class Γ is admissible if and only if its minimax level c is finite. But

$$c = \inf_{A \in \Gamma} \sup_{u \in A} J(u) = \inf_{\{u\} \in \Gamma} \sup_{u \in \{u\}} J(u) = \inf_{u \in H} J(u),$$

so that the class is admissible if and only if J is bounded from below.

The new, often very important information provided by the minimax principle, is the existence not only of a minimizing sequence, but of a *Palais–Smale* sequence at level $c = \inf_H J$. Thus, from now on, whenever a $C^{1,1}$ functional J is bounded from below, we can conclude that there exists a minimizing sequence $\{u_k\}_k$ *with the further noteworthy property* that

$$\nabla J(u_k) \to 0 \quad \text{in } H.$$

This property has a nontrivial proof even for real functions of one variable.

4.3 Two Classical Theorems

In this section we present the two most celebrated results of Critical Point Theory. We will see how they both fit into the framework given by the minimax principle. These two results have been generalized in all possible directions, but are still used very much in their original form.

Theorem 4.3.1 (Mountain Pass Theorem [5]) *Let H be a Hilbert space, and let $J \in C^{1,1}(H)$ satisfy $J(0) = 0$. Assume that there exist positive numbers ρ and α such that*

1. $J(u) \geq \alpha$ if $\|u\| = \rho$;
2. There exists $v \in H$ such that $\|v\| > \rho$ and $J(v) < \alpha$.

Then there exists a Palais–Smale sequence for J at a level $c \geq \alpha$. If J satisfies $(PS)_c$, then there exists a critical point at level c.

Proof We define the minimax class

$$\Gamma = \{\gamma([0,1]) \mid \gamma : [0,1] \to H \text{ is continuous}, \gamma(0) = 0, \gamma(1) = v\}$$

and the corresponding minimax level

$$c = \inf_{\gamma([0,1]) \in \Gamma} \sup_{u \in \gamma([0,1])} J(u) = \inf_{\gamma \in \Gamma} \max_{t \in [0,1]} J(\gamma(t)).$$

The sets in Γ are thus the continuous paths that connect the origin with the point v. In the last equality we identify γ and $\gamma([0,1])$, with some abuse of notation, as is customary to do.

If we prove that Γ is admissible, we apply the minimax principle, and the proof is done.

It is obvious that $c < +\infty$, since we maximize a continuous functional on a compact set. We have to prove that $c > -\infty$. To see this, we notice that every $\gamma \in \Gamma$ satisfies $\|\gamma(0)\| = 0$ and $\|\gamma(1)\| = \|v\| > \rho$. Since γ is continuous, there exists $t_\gamma \in [0,1]$ such that $\|\gamma(t_\gamma)\| = \rho$. This implies, by assumption 1,

$$\max_{t \in [0,1]} J(\gamma(t)) \geq J(\gamma(t_\gamma)) \geq \alpha.$$

As this holds for every $\gamma \in \Gamma$, we can take the infimum with respect to $\gamma \in \Gamma$ to obtain

$$c \geq \alpha > 0.$$

We now have to prove that for every $\varepsilon > 0$ small enough, Γ is invariant for all deformations that fix $J^{c-2\varepsilon}$.

For this, we take any $\varepsilon > 0$ so small that

$$J(0) < c - 2\varepsilon \quad \text{and} \quad J(v) < c - 2\varepsilon,$$

which is possible because $\max\{J(0), J(v)\} = \max\{0, J(v)\} < \alpha \leq c$.

Let now η be a deformation that fixes $J^{c-2\varepsilon}$ and take $\gamma \in \Gamma$. We have to prove that, for all $t \in [0,1]$,

$$\eta(t, \gamma([0,1])) \in \Gamma,$$

which means that $\eta(t, \gamma([0,1])) = \tilde{\gamma}([0,1])$ for some $\tilde{\gamma} \in \Gamma$. To this aim we define

$$\tilde{\gamma} : [0,1] \to H \quad \text{as } \tilde{\gamma}(s) = \eta(t, \gamma(s)).$$

Of course $\tilde{\gamma}$ is continuous. Since 0 and v belong to $J^{c-2\varepsilon}$, and η fixes this set, we have

$$\tilde{\gamma}(0) = \eta(t, \gamma(0)) = \eta(t, 0) = 0$$

and

$$\tilde{\gamma}(1) = \eta(t, \gamma(1)) = \eta(t, v) = v.$$

Hence $\tilde{\gamma}([0, 1]) \in \Gamma$, and Γ is an admissible class. We then apply Theorem 4.2.4 and we conclude the proof. $\qquad\square$

Some comments and remarks are in order.

Remark 4.3.2 The term "mountain pass" is justified by the geometrical properties of the graph of J. Indeed, it is customary to think of the points 0 and v as two villages. The assumptions of Theorem 4.3.1 imply that the two villages are separated by a mountain range: to go from 0 to v one must climb at least at a height α, which is strictly larger than the heights $J(0)$ and $J(v)$. The minimax class used in the proof is constructed exactly on this observation: one tries every possible road γ between 0 and v, measures the maximal height reached by the road ($\max_{t \in [0,1]} J(\gamma(t))$), and tries to minimize the maximal height. At the maximal height of the "optimal" road is located a (mountain) pass.

Remark 4.3.3 In the first part of this chapter, the attention was concentrated on the change of topology of different sublevels. In the Mountain Pass Theorem, this change is very simple to detect, and it is exactly what the assumptions describe: there is a sublevel which is not path-connected. Indeed, take a continuous path γ between 0 and v; this will lie in a sublevel J^b, where, for example $b = \max_{t \in [0,1]} J(\gamma(t))$. This means that 0 and v are in the same connected component of J^b. On the contrary, 0 and v are *not* in the same (arcwise) connected component of J^a, for every $a < c$. Thus J^b cannot be deformed into J^a.

Example 4.3.4 A functional that satisfies the assumptions of Theorem 4.3.1 is said to have a "mountain pass geometry". The functional associated to Problem (4.1), which we used as a motivation for this discussion, has a mountain pass geometry. Indeed, $J(0) = 0$, and one can take for example $\rho = C^{-\frac{p}{p-2}}$, $\alpha = (\frac{1}{2} - \frac{1}{p})C^{-\frac{2p}{p-2}}$, and v any function for which $\|v\| > \rho$ and $J(v)$ is negative; there are plenty of these functions since $J(\lambda u) \to -\infty$ as $\lambda \to +\infty$, for every $u \not\equiv 0$.

A large number of functionals J associated with superlinear problems have a mountain pass geometry. This is due to the fact that (i) 0 is a strict local minimum and J behaves well enough to prove that 1. in Theorem 4.3.1 is satisfied and (ii) $\lim_{\lambda \to +\infty} J(\lambda u) = -\infty$, which is guaranteed by superlinearity, implies 2.

Remark 4.3.5 As a final consideration, we wish to highlight the use of deformations that fix certain sets in the proof of Theorem 4.3.1. In view of the minimax principle, the proof of Theorem 4.3.1 reduces to the construction of an admissible minimax class. Now the class Γ is defined by three requirements: every γ is continuous and such that $\gamma(0) = 0$ and $\gamma(1) = v$. Of course every deformation, being continuous, preserves the continuity of γ, so the real condition to satisfy is the fact that the endpoints of γ should be held fixed during the deformations. We have obtained this by the use of deformations that fix a certain sublevel $J^{c-2\varepsilon}$ where the points 0 and v lie. Of course we do not want that these deformations fix also the sublevel J^c, otherwise the main argument of the minimax principle breaks down. Summing up, the reason why the class Γ is admissible comes from a *separation* property: the interesting level c is *strictly* larger than the levels that have to be kept fixed during the deformation, which are the levels $J(0) = 0$ and $J(v) < \alpha$. This aspect is present in almost every construction of concrete minimax classes. We will see another example of this kind in the Saddle Point Theorem below.

The Mountain Pass Theorem has been used an enormous amount of times in the literature to produce Palais–Smale sequence. Of course its use depends on the fact that the functional under study possesses the right geometry.

For functionals that do not have the mountain pass geometry, other minimax classes can be tried. One of the most popular results, that has also been generalized in many directions, is the Saddle Point Theorem. Roughly speaking, it works for functionals where the nonlinearity is not superlinear, but "asymptotically" linear at infinity. We postpone the examples to the next sections.

The proof of the Saddle Point Theorem requires a classical fixed point result, that we do not prove (for a proof see [2, 17], or [18]).

Theorem 4.3.6 (Brouwer)   *Let $B_R = \{x \in \mathbb{R}^N \mid |x| \le R\}$ and let $\psi : B_R \to B_R$ be a continuous function. Then ψ has a fixed point.*

We are going to use the following equivalent form of the Brouwer Theorem.

Theorem 4.3.7   *Let $B_R = \{x \in \mathbb{R}^N \mid |x| \le R\}$ and let $\varphi : B_R \to \mathbb{R}^N$ be a continuous function such that $\varphi(x) = x$ if $|x| = R$ (φ is the identity on ∂B_R). Then there exists $x \in B_R$ such that $\varphi(x) = 0$.*

Proof   We define a map $\psi : B_R \to B_R$ as

$$\psi(x) = \begin{cases} x - \varphi(x) & \text{if } |x - \varphi(x)| \le R, \\ \dfrac{R}{|x - \varphi(x)|}(x - \varphi(x)) & \text{if } |x - \varphi(x)| > R. \end{cases} \tag{4.10}$$

Clearly, the map ψ is well-defined, continuous and $\psi(B_R) \subseteq B_R$. By the Brouwer Theorem there exists $x \in B_R$ such that $\psi(x) = x$. If $|x - \varphi(x)| > R$, then

$$x = \frac{R}{|x - \varphi(x)|}(x - \varphi(x)),$$

and hence $|x| = R$. This implies $\varphi(x) = x$, so that $|x - \varphi(x)| = 0$, a contradiction. So it must be $|x - \varphi(x)| \le R$, and $x = \psi(x)$ then implies $x = x - \varphi(x)$, namely $\varphi(x) = 0$.   □

Remark 4.3.8   A linear n-dimensional subspace of a Hilbert space is isomorphic to $\mathbb{R}^n$. Therefore the preceding results holds in finite dimensional subspaces of Hilbert spaces. This is the form that we are going to use.

Theorem 4.3.9 (Saddle Point Theorem [42]) *Let H be a Hilbert space and assume that $J \in C^{1,1}(H)$. Let $E_n \subset H$ be a n-dimensional linear subspace of H and call $V \subset H$ its orthogonal complement in H, so that $H = E_n \oplus V$. For $R > 0$ let B_R^n be the closed ball in E_n with center at the origin and radius R, and let ∂B_R^n be its boundary in E_n. Assume that for some $R > 0$,*

$$\max_{u \in \partial B_R^n} J(u) < \inf_{u \in V} J(u). \tag{4.11}$$

Then there exists a Palais–Smale sequence for J at a level $c \geq \inf_V J$. If J satisfies $(PS)_c$, then there exists a critical point at level c.

Proof We define the minimax class

$$\Gamma = \left\{ \varphi(B_R^n) \mid \varphi : B_R^n \to H \text{ is continuous}, \varphi(u) = u \ \forall u \in \partial B_R^n \right\}$$

and the corresponding minimax level

$$c = \inf_{\varphi \in \Gamma} \max_{u \in B_R^n} J(\varphi(u)).$$

The sets in Γ are continuous n-dimensional surfaces of boundary $\partial B_R^n \subset E_n$.

As in Theorem 4.3.1, the proof amounts to showing that Γ is admissible. It is clear that $c < +\infty$; the interesting part is to check that $c > -\infty$.

Let $P : H \to E_n$ be the orthogonal projection on E_n, so that, $u \in V$ if and only if $P(u) = 0$ and $P(u) = u$ if and only if $u \in E_n$. Take a generic $\varphi \in \Gamma$ and define

$$\psi : B_R^n \to E_n \quad \text{as } \psi(u) = P(\varphi(u)).$$

The map ψ is of course continuous, and since φ is the identity on ∂B_R^n,

$$\text{for every } u \in \partial B_R^n, \quad \psi(u) = P(\varphi(u)) = P(u) = u.$$

So ψ is a continuous map from B_R^n to E_n that is the identity on ∂B_R^n. By Theorem 4.3.7, there exists $u_\varphi \in B_R^n$ such that $\psi(u_\varphi) = 0$, that is $P(\varphi(u_\varphi)) = 0$, which means $\varphi(u_\varphi) \in V$. Therefore,

$$\max_{u \in B_R^n} J(\varphi(u)) \geq J(\varphi(u_\varphi)) \geq \inf_{u \in V} J(u) > \max_{u \in \partial B_R^n} J(u) > -\infty.$$

As this holds for every $\varphi \in \Gamma$, taking the infimum over Γ yields

$$c \geq \inf_{u \in V} J(u) > \max_{u \in \partial B_R^n} J(u) > -\infty.$$

Now we have to prove that Γ satisfies the invariance properties of Definition 4.2.3. Let $\varepsilon > 0$ be any number so small that

$$\inf_V J - 2\varepsilon > \max_{\partial B_R^n} J,$$

and take a deformation η that fixes $J^{c-2\varepsilon}$.

We want to prove that if $A = \varphi(B_R^n) \in \Gamma$, then $\eta(t, A) \in \Gamma$ for all $t \in [0, 1]$. To this aim we define

$$\overline{\varphi} : B_R^n \to H \quad \text{as} \quad \overline{\varphi}(u) = \eta(t, \varphi(u)).$$

Of course $\overline{\varphi}$ is continuous. We have to check that $\overline{\varphi}(u) = u$ for all $u \in \partial B_R^n$. For this, take $u \in \partial B_R^n$. By definition, $\varphi(u) = u$, and since

$$J(u) \leq \max_{\partial B_R^n} J < \inf_V J - 2\varepsilon \leq c - 2\varepsilon,$$

$\eta(t, u) = u$ for all t. Therefore

$$\overline{\varphi}(u) = \eta(t, \varphi(u)) = \eta(t, u) = u,$$

which shows that $\overline{\varphi}$ is the identity on ∂B_R^n. The class Γ is admissible and the application of Theorem 4.2.4 concludes the proof. $\square$

Remark 4.3.10 The preceding proof is identical in its structure to the proof of the Mountain Pass Theorem. In particular, also in the Saddle Point Theorem the key property is the *separation* of the minimax level from the levels that have to be kept fixed by the deformations. This is the scope of assumption (4.11): the minimax level c is greater than or equal to $\inf_V J$, which is *strictly* greater than $\alpha = \max_{\partial B_R^n} J$. Thus we can construct a deformation that fixes J^α, where ∂B_R^n lies, and this is ultimately what makes Γ admissible.

4.4 Some Applications

We now illustrate the use of the previous theorems by some applications. We have chosen the simplest cases, but the reader should keep in mind that a very large number of problems coming from physics, mechanics or geometry can be dealt with the two results of the preceding section.

The structure of the following results is suggested by that of the Mountain Pass Theorem and of the Saddle Point Theorem: one checks that the Palais–Smale condition holds, and, separately, that the "geometry" of the functionals involved fits into the abstract requirements of the mentioned theorems.

4.4.1 Superlinear Problems

We begin with some superlinear nonlinearities, for which the Mountain Pass Theorem is tailored.

Consider the following problem: given a bounded open set $\Omega \subset \mathbb{R}^N$, $N \geq 3$, and a nonnegative L^∞ function q defined on Ω, find u such that

$$\begin{cases} -\Delta u + q(x)u = f(u) & \text{in } \Omega, \\ u = 0 & \text{on } \partial\Omega. \end{cases} \tag{4.12}$$

A problem of this form has already been discussed in Sect. 2.5. We will now weaken the assumptions, maintaining the validity of the result.

We set $F(t) = \int_0^t f(s)\, ds$ and we consider the following set of assumptions.

(**f₆**) There exist $p \in (2, 2^*)$ and $C > 0$ such that for all $s, t \in \mathbb{R}$,

$$|f(t) - f(s)| \leq C|t - s|\,(|s| + |t| + 1)^{p-2}.$$

(**f₇**) $\lim_{t \to 0} \frac{f(t)}{t} = 0$.

(**f₈**) There exist $M > 0$ and $\mu > 2$ such that $f(t)t \geq \mu F(t)$ when $|t| \geq M$.

(**f₉**) There exists $t_0 \in \mathbb{R}$ with $|t_0| \geq M$ such that $F(t_0) > 0$.

Remark 4.4.1 The meaning of assumption (**f₆**) is twofold. First of all, it is a growth condition: setting $s = 0$ one sees that (**f₆**) implies

$$|f(t)| \leq K(|t|^{p-1} + 1) \tag{4.13}$$

for all t and for a suitable positive constant K.

Next, (**f₆**) implies that f is locally Lipschitz continuous on $\mathbb{R}$. These two facts allow one to prove that the functional $G : H_0^1(\Omega) \to \mathbb{R}$ defined by

$$G(u) = \int_\Omega F(u)\,dx$$

is of class $C^{1,1}$ and $G'(u)v = \int_\Omega f(u)v\,dx$ for all $v \in H_0^1(\Omega)$.

Also, assumption (**f₈**) means that the function $t \mapsto \frac{F(t)}{|t|^\mu}$ is increasing for $t \geq M$ and decreasing for $t \leq -M$, as one can immediately see by differentiation.

Remark 4.4.2 Problem (4.1), which we used as a motivation for the construction of the minimax methods, is a particular case of Problem (4.12).

If, as usual, we equip $H_0^1(\Omega)$ with the norm

$$\|u\|^2 = \int_\Omega |\nabla u|^2 + q(x)u^2\,dx, \tag{4.14}$$

we can say that weak solutions of Problem (4.12) are critical points of the functional $J : H_0^1(\Omega) \to \mathbb{R}$ defined as

$$J(u) = \frac{1}{2}\int_\Omega |\nabla u|^2 + q(x)u^2\,dx - \int_\Omega F(u)\,dx = \frac{1}{2}\|u\|^2 - G(u),$$

which is in $C^1(H_0^1(\Omega))$.

We will prove the following theorem, where assumption (**h₁**) is the one introduced at the beginning of Sect. 2.1 (recall also Remark 2.1.1).

Theorem 4.4.3 *Assume that* (**h₁**) *and* (**f₆**)–(**f₉**) *hold. Then Problem (4.12) has at least one non trivial solution.*

In view of Remark 4.1.16, the abstract minimax principle of Sect. 4.2 holds by merely assuming J of class C^1, and so do the Mountain Pass and Saddle Point theorems. Therefore, in the previous result, assumption (**f₆**) could be replaced by the continuity of f and (4.13).

Remark 4.4.4 Theorem 4.4.3 gives a stronger result than those we proved in Sect. 2.5. Indeed, it is easy to see that hypotheses $(\mathbf{f_1})$–$(\mathbf{f_4})$ imply $(\mathbf{f_6})$–$(\mathbf{f_9})$ (with $t_0 > M$); therefore Theorem 2.5.2 is a special case of Theorem 4.4.3, except for the positivity of the solution. However, if $t_0 > M$ this can be recovered by the method of Example 1.7.10 after extending f to zero for negative arguments. Moreover, the minimax method allows one to drop the hypothesis of C^1 nonlinearities and the assumption $(\mathbf{f_4})$. Also the hypotheses of Theorem 2.5.11 imply $(\mathbf{f_6})$–$(\mathbf{f_9})$.

We are going to show that the Mountain Pass Theorem is applicable. First, a technical result.

Lemma 4.4.5 *The functional J is in* $C^{1,1}(H_0^1(\Omega))$.

Proof We already know that J is C^1; of course, since the differential of the quadratic term is (globally) Lipschitz continuous, we only have to check the local Lipschitz continuity for the differential of the functional $G(u)$.

To this aim, fix $R > 0$ and take $u_1, u_2 \in H_0^1(\Omega)$ such that $\|u_1\|, \|u_2\| \le R$. Then, for all $v \in H_0^1(\Omega)$ we have, by $(\mathbf{f_6})$,

$$
|(G'(u_1) - G'(u_2))v| = \left| \int_\Omega (f(u_1) - f(u_2))v\, dx \right|
$$

$$
\le \left(\int_\Omega |f(u_1) - f(u_2)|^{\frac{p}{p-1}}\, dx \right)^{\frac{p-1}{p}} \left(\int_\Omega |v|^p\, dx \right)^{1/p}
$$

$$
\le C \left(\int_\Omega |u_1 - u_2|^{\frac{p}{p-1}} (1 + |u_1| + |u_2|)^{\frac{p(p-2)}{p-1}}\, dx \right)^{\frac{p-1}{p}} \|v\|.
$$

Applying the Hölder inequality with exponents $p - 1$ and $\frac{p-1}{p-2}$ we obtain

$$
\int_\Omega |u_1 - u_2|^{\frac{p}{p-1}} (1 + |u_1| + |u_2|)^{\frac{p(p-2)}{p-1}}\, dx
$$

$$
\le \left(\int_\Omega |u_1 - u_2|^p\, dx \right)^{\frac{1}{p-1}} \left(\int_\Omega (1 + |u_1| + |u_2|)^p\, dx \right)^{\frac{p-2}{p-1}}
$$

$$
\le \|u_1 - u_2\|^{\frac{p}{p-1}} K(R),
$$

where $K(R)$ depends on R but not on u_1, u_2. This shows that

$$
|(G'(u_1) - G'(u_2))v| \le \|u_1 - u_2\| K(R)^{\frac{p-1}{p}} \|v\|,
$$

and therefore, maximizing over all $\|v\| = 1$,

$$
\|G'(u_1) - G'(u_2)\| \le C(R)\|u_1 - u_2\|,
$$

where $C(R)$ is a constant depending on R. So we have proved that G' (and hence J) is locally Lipschitz continuous. $\qquad\square$

Now we turn to the compactness properties of J.

Lemma 4.4.6 *The functional J satisfies* $(PS)_c$ *for every* $c \in \mathbb{R}$.

Proof Take $c \in \mathbb{R}$ and assume that $\{u_k\}$ is a Palais–Smale sequence at level c, namely such that

$$J(u_k) \to c \quad \text{and} \quad J'(u_k) \to 0 \quad (\text{in } (H_0^1(\Omega))').$$

This implies of course that there is a constant $C > 0$ such that

$$|J(u_k)| \le C \quad \text{and} \quad |J'(u_k)u_k| \le C\|u_k\| \tag{4.15}$$

for every k. The following is a standard trick in superlinear problems to show that Palais–Smale sequences are bounded, and is the main motivation for the use of assumption $(\mathbf{f_8})$. By (4.15) we can write

$$C(1 + \|u_k\|) \ge J(u_k) - \frac{1}{\mu} J'(u_k)u_k$$

$$= \left(\frac{1}{2} - \frac{1}{\mu}\right)\|u_k\|^2 + \int_\Omega \left(\frac{1}{\mu} f(u_k)u_k - F(u_k)\right) dx.$$

Then from $(\mathbf{f_8})$ we obtain

$$\int_\Omega \left(\frac{1}{\mu} f(u_k)u_k - F(u_k)\right) dx = \int_{\{|u_k| \le M\}} \left(\frac{1}{\mu} f(u_k)u_k - F(u_k)\right) dx$$

$$+ \int_{\{|u_k| > M\}} \left(\frac{1}{\mu} f(u_k)u_k - F(u_k)\right) dx$$

$$\ge \int_{\{|u_k| \le M\}} \left(\frac{1}{\mu} f(u_k)u_k - F(u_k)\right) dx \ge -C_1,$$

and hence

$$C(1 + \|u_k\|) \ge \left(\frac{1}{2} - \frac{1}{\mu}\right)\|u_k\|^2 - C_1,$$

which of course implies that $\{u_k\}$ is bounded.

By usual arguments we can assume that, up to a subsequence, there exists $u \in H_0^1(\Omega)$ such that

- $u_k \rightharpoonup u$ in $H_0^1(\Omega)$,
- $u_k \to u$ in $L^q(\Omega)$ for all $q \in [2, 2^*)$,
- $u_k(x) \to u(x)$ for almost every $x \in \Omega$.

We now show that the convergence of u_k to u is strong, and this is a consequence of the fact that $p < 2^*$.

First of all, from the above convergence properties, we obtain by usual arguments,

$$\int_\Omega f(u_k)u_k \, dx \to \int_\Omega f(u)u \, dx \quad \text{and} \quad \int_\Omega f(u_k)u \, dx \to \int_\Omega f(u)u \, dx. \tag{4.16}$$

As $J'(u_k) \to 0$ and $u_k \rightharpoonup u$, we also have $J'(u_k)(u_k - u) \to 0$ and obviously $J'(u)(u_k - u) \to 0$. Then, as $k \to \infty$,

$$o(1) = \big(J'(u_k) - J'(u)\big)(u_k - u) = \|u_k - u\|^2 - \int_\Omega \big(f(u_k) - f(u)\big)(u_k - u)\,dx$$

$$= \|u_k - u\|^2 + o(1)$$

by (4.16). This shows that $u_k \to u$ in $H_0^1(\Omega)$, and proves that J satisfies $(PS)_c$ for every $c \in \mathbb{R}$. $\square$

End of the proof of Theorem 4.4.3 We just have to check that the geometrical assumptions of the Mountain Pass Theorem are satisfied.

First of all, as $F(0) = 0$, we have $J(0) = 0$. We now prove that hypothesis 1 of Theorem 4.3.1 holds.

Let $\lambda_1 > 0$ be the first eigenvalue of the operator $-\Delta + q$ on Ω. Fix $\varepsilon > 0$ so small that

$$\frac{1}{2} - \frac{\varepsilon}{\lambda_1} \geq \frac{1}{4}.$$

By ($\mathbf{f_7}$) we have $\lim_{t \to 0} \frac{F(t)}{t^2} = 0$, and therefore there exists $\delta > 0$ such that

$$F(t) \leq \varepsilon t^2$$

if $|t| \leq \delta$. By ($\mathbf{f_6}$), as we remarked above, we deduce that there are $a, b > 0$ such that $|f(t)| \leq a + b|t|^{2^*-1}$, and then also

$$|F(t)| \leq \tilde{a} + \tilde{b}|t|^{2^*}$$

for some positive $\tilde{a}, \tilde{b}$ and all $t \in \mathbb{R}$. Collecting these inequalities, it is not difficult to see that there is a constant $C_\varepsilon > 0$ such that

$$|F(t)| \leq \varepsilon t^2 + C_\varepsilon |t|^{2^*}$$

for all t. Hence

$$J(u) = \frac{1}{2}\|u\|^2 - \int_\Omega F(u)\,dx \geq \frac{1}{2}\|u\|^2 - \varepsilon \int_\Omega |u|^2\,dx - C_\varepsilon \int_\Omega |u|^{2^*}\,dx$$

$$\geq \frac{1}{2}\|u\|^2 - \frac{\varepsilon}{\lambda_1}\|u\|^2 - C\|u\|^{2^*} \geq \frac{1}{4}\|u\|^2 - C\|u\|^{2^*},$$

from which the assumption 1. of Theorem 4.3.1 easily follows, for example taking as ρ any number so small that $\frac{1}{4}\rho^2 - C\rho^{2^*}$ is positive.

To check that assumption 2 also holds, we notice that it is easy, using the same arguments of Lemma 2.5.3, to show that there is a constant $C > 0$ such that $F(t) \geq C|t|^\mu$ for all $t \geq t_0$, if $t_0 > M$, and for all $t \leq t_0$, if $t_0 < -M$. To proceed, we assume that $t_0 \geq M$ (the other case being similar). Take $\varphi \in C_0^\infty(\Omega)$ such that $\varphi(x) \geq 0$ for all $x \in \Omega$ and $\varphi(x) \geq 1$ on some ball $B = B_r(x_0) \subset \Omega$. Then

$$t\varphi(x) \geq t_0, \quad \text{and hence} \quad F(t\varphi(x)) \geq Ct^\mu \varphi^\mu(x) \quad \text{for all } t \geq t_0 \text{ and all } x \in B,$$

while $|F(t\varphi)| \leq C$ if $0 \leq t\varphi(x) \leq t_0$. So we obtain

$$\int_\Omega F(t\varphi)\,dx = \int_{\{0\le t\varphi\le t_0\}} F(t\varphi)\,dx + \int_{\{t\varphi>t_0\}} F(t\varphi)\,dx$$

$$\ge -C + C\int_{\{t\varphi>t_0\}} t^\mu \varphi^\mu\,dx$$

$$\ge -C + C\int_B t^\mu \varphi^\mu\,dx \ge -C + Ct^\mu,$$

and then

$$J(t\varphi) = \frac{t^2}{2}\|\varphi\|^2 - \int_\Omega F(t\varphi)\,dx \le t^2 C_1 + C - Ct^\mu.$$

As $\mu > 2$, it is now obvious that for large t we have $J(t\varphi) < 0$, so also assumption 2 is proved. We then apply Theorem 4.3.1 to obtain a critical point at level c. As $c > 0$, this critical point cannot be zero, and therefore the solution of Problem (4.12) is nontrivial. $\qquad\square$

We now give another application of the Mountain Pass Theorem. On a bounded open set $\Omega \subset \mathbb{R}^N$, we look for a solution u of

$$\begin{cases} -\Delta u + q(x)u = \lambda u + |u|^{p-2}u & \text{in } \Omega, \\ u = 0 & \text{on } \partial\Omega, \end{cases} \tag{4.17}$$

with $\lambda \in \mathbb{R}$. As usual, q is bounded and nonnegative on Ω, and we assume $2 < p < 2^*$. Weak solutions of (4.17) are critical points of the differentiable functional $J : H_0^1(\Omega) \to \mathbb{R}$ defined by

$$J(u) = \frac{1}{2}\|u\|^2 - \frac{\lambda}{2}\int_\Omega u^2\,dx - \frac{1}{p}\int_\Omega |u|^p\,dx,$$

where the norm of u is the one defined in (4.14).

Remark 4.4.7 It is easy to see that if we define $f(u) = \lambda u + |u|^{p-2}u$, then f satisfies $(\mathbf{f_6})$, $(\mathbf{f_8})$, $(\mathbf{f_9})$, but not $(\mathbf{f_7})$. Therefore we cannot use directly Theorem 4.4.3 to solve Problem (4.17). Actually one could reenter in the frame of Theorem 4.4.3 by assuming that $\lambda_1(-\Delta + q(x) - \lambda) > 0$ and use as an equivalent norm in $H_0^1(\Omega)$ the first two terms of J. We prefer not do so because in the applications one encounters frequently problems written as (4.17).

We are going to prove the following result; in its statement we denote by $\lambda_1(q)$ the first eigenvalue of the operator $-\Delta + q(x)$.

Theorem 4.4.8 *Assume that* $(\mathbf{h_1})$ *holds and that* $2 < p < 2^*$. *If* $\lambda < \lambda_1(q)$, *then Problem* (4.17) *has at least one non trivial solution.*

Proof The argument is essentially the same as that of the previous theorem. The nonlinearity $f(u) = \lambda u + |u|^{p-2}u$ in (4.17) is C^1 and satisfies $(\mathbf{f_6})$, so the proof of Lemma 4.4.5 works in the same way and we obtain that $J \in C^{1,1}(H_0^1(\Omega))$.

Let us check that the Palais–Smale condition holds at every level. Let $\{u_k\}$ be a sequence such that

$$J(u_k) \to c \quad \text{and} \quad J'(u_k) \to 0.$$

As in the previous theorem, this implies that there is a constant $C > 0$ such that

$$C(1 + \|u_k\|) \geq J(u_k) - \frac{1}{p}J'(u_k)u_k$$

$$= \left(\frac{1}{2} - \frac{1}{p}\right)\|u_k\|^2 - \left(\frac{\lambda}{2} - \frac{\lambda}{p}\right)\int_\Omega u_k^2 \, dx - \frac{1}{p}\int_\Omega |u_k|^p \, dx$$

$$+ \frac{1}{p}\int_\Omega |u_k|^p \, dx$$

$$= \left(\frac{1}{2} - \frac{1}{p}\right)\|u_k\|^2 - \lambda\left(\frac{1}{2} - \frac{1}{p}\right)\int_\Omega u_k^2 \, dx$$

$$\geq \left(\frac{1}{2} - \frac{1}{p}\right)\|u_k\|^2 - \frac{\lambda}{\lambda_1(q)}\left(\frac{1}{2} - \frac{1}{p}\right)\|u_k\|^2$$

$$= \left(\frac{1}{2} - \frac{1}{p}\right)\left(1 - \frac{\lambda}{\lambda_1(q)}\right)\|u_k\|^2.$$

Since $\lambda < \lambda_1(q)$, this implies that $\{u_k\}$ is bounded and, as in the previous theorem, we deduce that there is $u \in H_0^1(\Omega)$ such that

- $u_k \rightharpoonup u$ in $H_0^1(\Omega)$,
- $u_k \to u$ in $L^q(\Omega)$ for all $q \in [2, 2^*)$,
- $u_k(x) \to u(x)$ for almost all $x \in \Omega$.

Then we easily see that as $k \to \infty$,

$$\int_\Omega |u_k|^{p-2}u_k(u_k - u) \, dx = o(1).$$

We conclude by computing

$$o(1) = (J'(u_k) - J'(u))(u_k - u)$$

$$= \|u_k - u\|^2 - \lambda\int_\Omega |u_k - u|^2 \, dx - \int_\Omega \left(|u_k|^{p-2}u_k - |u|^{p-2}u\right)(u_k - u) \, dx$$

$$= \|u_k - u\|^2 + o(1),$$

which shows that $u_k \to u$ strongly. Condition $(PS)_c$ holds for every c.

Finally we check that the geometric assumptions of Theorem 4.3.1 are satisfied. It is obvious that $J(0) = 0$. For every $u \in H_0^1(\Omega)$ we have

$$J(u) = \frac{1}{2}\|u\|^2 - \frac{\lambda}{2}\int_\Omega u^2 \, dx - \frac{1}{p}\int_\Omega |u|^p \, dx$$

$$\geq \frac{1}{2}\|u\|^2 - \frac{\lambda}{2\lambda_1(q)}\|u\|^2 - C\|u\|^p = \frac{1}{2}\left(1 - \frac{\lambda}{\lambda_1(q)}\right)\|u\|^2 - C\|u\|^p.$$

As $\lambda < \lambda_1(q)$ and $p > 2$, it is enough to choose any positive number ρ so small that $\frac{1}{2}(1 - \frac{\lambda}{\lambda_1(q)})\rho^2 - C\rho^p > 0$. Hypothesis 1 of Theorem 4.3.1 is satisfied.

Finally for every $u \in H_0^1(\Omega) \setminus \{0\}$ and $t > 0$ we have

$$J(tu) = \frac{t^2}{2}\|u\|^2 - \frac{\lambda t^2}{2}\int_\Omega u^2\,dx - \frac{t^p}{p}\int_\Omega |u|^p\,dx$$

$$= \frac{t^2}{2}\left(\|u\|^2 - \lambda\int_\Omega u^2\,dx\right) - \frac{t^p}{p}\int_\Omega |u|^p\,dx.$$

Therefore $J(tu) \to -\infty$ as $t \to +\infty$, and we can certainly find v such that $\|v\| \geq \rho$ and $J(v) < 0$. The mountain Pass Theorem then yields the existence of a critical point u for J at a positive level. This critical point is then a nontrivial solution of Problem (4.17). $\qquad\square$

Remark 4.4.9 If $\lambda \geq \lambda_1(q)$, the Mountain Pass Theorem cannot be applied, because the "geometry around zero" is lost, in the sense that zero is no longer a local minimum for J. To see this, let φ_1 be the eigenfunction associated to $\lambda_1(q)$. For every $t \neq 0$ we have

$$J(t\varphi_1) = \frac{t^2}{2}\|\varphi_1\|^2 - \frac{\lambda t^2}{2}\int_\Omega \varphi_1^2\,dx - \frac{|t|^p}{p}\int_\Omega |\varphi_1|^p\,dx$$

$$\leq \frac{t^2}{2}\|\varphi_1\|^2 - \frac{\lambda_1(q)t^2}{2}\int_\Omega \varphi_1^2\,dx - \frac{|t|^p}{p}\int_\Omega |\varphi_1|^p\,dx$$

$$= -\frac{|t|^p}{p}\int_\Omega |\varphi_1|^p\,dx < 0,$$

and J is strictly negative on all the line $t\varphi_1$, except $t = 0$.

4.4.2 Asymptotically Linear Problems

We now present some applications of the Saddle Point Theorem. The most frequent situation where this theorem is applicable is that of asymptotically linear problems, namely those where the nonlinearity grows linearly at infinity.

To illustrate the method, we consider an open and bounded subset Ω of $\mathbb{R}^N$, and we look for a solution of

$$\begin{cases} -\Delta u + q(x)u = \lambda u + f(u) + h(x) & \text{in } \Omega, \\ u = 0 & \text{on } \partial\Omega. \end{cases} \tag{4.18}$$

Throughout this section, in addition to $(\mathbf{h_1})$ and $(\mathbf{h_2})$, we will use the following assumption:

$(\mathbf{f_{10}})$ $f \in C^1(\mathbb{R})$, f is bounded and $|f'(t)| \leq C(1 + |t|^{2^*-2})$.

The requirement that f be C^1 can be weakened, and we use it only to avoid some technicalities (for example we could take f bounded and satisfying $(\mathbf{f_6})$). If we define

$$\tilde{f} : \Omega \times \mathbb{R} \to \mathbb{R} \quad \text{as} \quad \tilde{f}(x,t) = \lambda t + f(t) + h(x),$$

we see that

$$\lim_{t \to \pm\infty} \frac{\tilde{f}(t,x)}{t} = \lambda$$

almost everywhere in Ω. Thus $\tilde{f}$ grows linearly in t at infinity, and this characterizes (4.18) as an asymptotically linear problem.

Once $H_0^1(\Omega)$ is equipped with the usual norm defined by (4.14), the weak solutions of (4.18) are the critical points of the functional $J : H_0^1(\Omega) \to \mathbb{R}$ defined by

$$J(u) = \frac{1}{2} \|u\|^2 - \frac{\lambda}{2} \int_\Omega u^2 \, dx - \int_\Omega F(u) \, dx - \int_\Omega hu \, dx,$$

where of course $F(t) = \int_0^t f(s) \, ds$. Due to $(\mathbf{f_{10}})$, repeating the arguments of Lemma 4.4.5, the functional J is readily seen to be in $C^{1,1}(H_0^1(\Omega))$, and we will not discuss this point anymore.

Let $0 < \lambda_1 < \lambda_2 \le \cdots$ be the usual sequence of the eigenvalues of the operator $-\Delta + q(x)$. The solvability of Problem (4.18) strongly depends on the location of λ with respect to the eigenvalues λ_k.

Definition 4.4.10 If $\lambda \ne \lambda_k$ for every k, Problem (4.18) is said to be nonresonant. If $\lambda = \lambda_k$ for some k, the problem is said to be resonant or at resonance with the k-th eigenvalue.

Resonant problems are much more difficult to solve, and it may happen that they do not have any solution at all, even in the linear case (compare with Theorem 1.7.8). So, to illustrate the method we first consider a nonresonant problem.

Theorem 4.4.11 *Assume that* $(\mathbf{h_1})$, $(\mathbf{h_2})$ *and* $(\mathbf{f_{10}})$ *hold. If* $\lambda \ne \lambda_k$ *for every k, then Problem* (4.18) *has at least one solution.*

If $\lambda < \lambda_1$, the functional J is coercive and can be shown to have a global minimum, with the methods of Sect. 2.1. The interesting case is thus $\lambda > \lambda_1$, and we define $n \in \mathbb{N}$ to be the number such that

$$\lambda_n < \lambda < \lambda_{n+1}. \tag{4.19}$$

In this case we are going to show that the functional J satisfies the assumptions of the Saddle Point Theorem. As usual we split the problem in two parts: compactness and geometric assumptions.

In what follows, we always take for granted the assumptions of Theorem 4.4.11 and (4.19).

Lemma 4.4.12 *The functional J satisfies* $(PS)_c$ *for every* $c \in \mathbb{R}$.

Proof Let $\{u_k\}$ be a sequence such that

$$J(u_k) \to c \quad \text{and} \quad J'(u_k) \to 0.$$

We first prove that $\{u_k\}$ is bounded.

Assume for contradiction that, up to a subsequence, $\|u_k\| \to +\infty$. Since $J'(u_k) \to 0$, for every $v \in H_0^1(\Omega)$ we can write $J'(u_k)v = o(1)\|v\|$, namely

$$(u_k|v) - \lambda \int_\Omega u_k v \, dx - \int_\Omega f(u_k) v \, dx - \int_\Omega h v \, dx = o(1)\|v\|. \quad (4.20)$$

Define $\psi_k = \frac{u_k}{\|u_k\|}$. Of course the sequence $\{\psi_k\}$ is bounded in $H_0^1(\Omega)$, so we can assume $\psi_k \rightharpoonup \psi \in H_0^1(\Omega)$. Dividing by $\|u_k\|$ the equation above we obtain

$$(\psi_k|v) - \lambda \int_\Omega \psi_k v \, dx - \frac{1}{\|u_k\|} \int_\Omega f(u_k) v \, dx - \frac{1}{\|u_k\|} \int_\Omega h v \, dx = o(1)\frac{\|v\|}{\|u_k\|}.$$

As f is bounded, passing to the limit we get

$$(\psi|v) - \lambda \int_\Omega \psi v \, dx = 0. \quad (4.21)$$

Let us now prove that $\psi \not\equiv 0$. Assume instead that $\psi = 0$. If this is the case, by $J'(u_k)\psi_k = o(1)$ we first obtain

$$(u_k|\psi_k) - \lambda \int_\Omega u_k \psi_k \, dx - \int_\Omega f(u_k)\psi_k \, dx - \int_\Omega h\psi_k \, dx = o(1),$$

and then, dividing by $\|u_k\|$,

$$\|\psi_k\|^2 - \lambda \int_\Omega \psi_k^2 \, dx = o(1).$$

We know that $\psi_k \rightharpoonup 0$ in $H_0^1(\Omega)$, which implies $\psi_k \to 0$ in $L^2(\Omega)$. Therefore we obtain $\|\psi_k\| \to 0$, a contradiction. So we have constructed a non zero function ψ satisfying (4.21) for all $v \in H_0^1(\Omega)$, namely a weak solution of

$$\begin{cases} -\Delta\psi + q(x)\psi = \lambda\psi & \text{in } \Omega, \\ \psi = 0 & \text{on } \partial\Omega. \end{cases} \quad (4.22)$$

But this says that λ is an eigenvalue of $-\Delta + q(x)$, which contradicts the nonresonance assumption.

This proves that $\{u_k\}$ is a bounded sequence, so we can assume $u_k \rightharpoonup u$ in $H_0^1(\Omega)$ and $u_k \to u$ in $L^2(\Omega)$. From this we obtain easily with the usual arguments

$$\begin{aligned} o(1) &= \left(J'(u_k) - J'(u)\right)(u_k - u) \\ &= \|u_k - u\|^2 - \lambda \int_\Omega (u_k - u)^2 \, dx - \int_\Omega (f(u_k) - f(u))(u_k - u) \, dx \\ &\quad - \int_\Omega h(u_k - u) \\ &= \|u_k - u\|^2 + o(1), \end{aligned}$$

which shows that $u_k \to u$ in H_0^1; the functional J satisfies $(PS)_c$ for all $c \in \mathbb{R}$. $\quad\square$

End of the proof of Theorem 4.4.11 We just have to check that J satisfies the geometric assumption of the Saddle Point Theorem. Let $\{\varphi_k\}_k$ be the sequence of the eigenfunctions of $-\Delta + q(x)$ (orthonormal in $L^2(\Omega)$), each associated to an eigenvalue λ_k. Define

$$E_n = \text{span}\{\varphi_1, \ldots, \varphi_n\}, \quad \text{and} \quad V = E_n^\perp = \text{cl}\{\text{span}\{\varphi_k \mid k \geq n + 1\}\}.$$

These two subspaces are nothing else than the spaces X_1 and X_2 introduced in (2.8). We recall from Lemma 2.2.5 that

$$\|u\|^2 \leq \lambda_n |u|_2^2 \quad \forall u \in E_n \quad \text{and} \quad \|u\|^2 \geq \lambda_{n+1} |u|_2^2 \quad \forall u \in V. \quad (4.23)$$

From this we deduce, for every $u \in E_n$,

$$J(u) = \frac{1}{2}\|u\|^2 - \frac{\lambda}{2}|u|_2^2 - \int_\Omega F(u)\,dx - \int_\Omega hu\,dx$$

$$\leq \frac{1}{2}\|u\|^2 - \frac{\lambda}{2\lambda_n}\|u\|^2 + C\|u\| = \frac{1}{2}\left(1 - \frac{\lambda}{\lambda_n}\right)\|u\|^2 + C\|u\|.$$

Denoting by B_R^n the ball in E_n of center zero and radius R, the last inequality shows that for every $u \in \partial B_R^n$,

$$J(u) \leq \frac{1}{2}\left(1 - \frac{\lambda}{\lambda_n}\right)R^2 + CR.$$

Since $\lambda/\lambda_n > 1$, we have proved that

$$\max_{u \in \partial B_R^n} J(u) \to -\infty$$

if $R \to \infty$.

We now prove that J is bounded from below on V. Still from (4.23), for every $u \in V$ we have

$$J(u) = \frac{1}{2}\|u\|^2 - \frac{\lambda}{2}|u|_2^2 - \int_\Omega F(u)\,dx - \int_\Omega hu\,dx$$

$$\geq \frac{1}{2}\|u\|^2 - \frac{\lambda}{2\lambda_{n+1}}\|u\|^2 - C\|u\|$$

$$= \frac{1}{2}\left(1 - \frac{\lambda}{\lambda_{n+1}}\right)\|u\|^2 - C\|u\|.$$

Since $\lambda/\lambda_{n+1} < 1$, this implies $\inf_V J > -\infty$. Hence for R large enough we can guarantee that

$$\max_{u \in \partial B_R^n} J(u) < \inf_{u \in V} J(u).$$

All the assumptions of the Saddle Point Theorem are satisfied and we apply it to conclude.   □

Remark 4.4.13 It is interesting to compare this result with the similar one of Sect. 2.2. In Theorem 2.2.3 a *global* assumption on the derivative of f was made, namely (2.7). In Theorem 4.4.11 there is *no assumption* on f', except of course a natural requirement on its growth.

Example 4.4.14 Whenever f and f' are bounded on $\mathbb{R}$ (and λ is not an eigenvalue), the assumptions of Theorem 4.4.11 are satisfied. Thus, on an open and bounded $\Omega \subset \mathbb{R}^N$ the homogeneous Dirichlet problem for equations like

$$-\Delta u + q(x)u = \lambda u + \arctan(u) + h(x)$$

or

$$-\Delta u + q(x)u = \lambda u + \frac{u}{1+u^2} + h(x)$$

always admits a solution, for every $h \in L^2(\Omega)$.

4.4.3 A Problem at Resonance

The Saddle Point Theorem is still applicable for certain problems of type (4.18) in presence of *resonance*, namely when λ coincides with one of the eigenvalues of $-\Delta + q$. These problems are generally harder to solve, and one does not expect solutions without imposing further conditions on the data of the problem.

In the literature these assumptions are normally expressed as follows: one *imposes* some conditions on f, which determine the type of nonlinearity under study, and then tries to understand for which types of h the problem is solvable.

In this section we consider, on a bounded open set Ω, the problem

$$\begin{cases} -\Delta u = \lambda u + f(u) + h(x) & \text{in } \Omega, \\ u = 0 & \text{on } \partial\Omega. \end{cases} \tag{4.24}$$

We assume here that $q \equiv 0$, to simplify notation, but everything would work just as well replacing $-\Delta$ with $-\Delta + q(x)$, with q satisfying (**h₁**) (possibly replacing the sign condition by (2.3)).

The characteristic features of the problem are described by the following set of assumptions. The function f satisfies (**f₁₀**) and the limits

$$f_l = \lim_{t \to -\infty} f(t) \quad \text{and} \quad f_r = \lim_{t \to +\infty} f(t)$$

exist, are finite and are *different*; to fix ideas we suppose that

$$f_l > f_r, \tag{4.25}$$

but the other case would work as well, with some modifications in the main argument. This specifies a *class* of nonlinearities.

Remark 4.4.15 Problem (4.24) can be seen as a nonlinear generalization of a non-homogeneous linear problem, such as

$$\begin{cases} -\Delta u = \lambda u + h(x) & \text{in } \Omega, \\ u = 0 & \text{on } \partial\Omega \end{cases} \tag{4.26}$$

discussed in Sect. 1.7. When λ is an eigenvalue of the Laplacian, Theorem 1.7.8 says that the existence of solutions is guaranteed *provided* h is orthogonal to the eigenspace relative to λ. Adding a nonlinear term $f(u)$ in the equation obviously alters significantly the simple structure of the linear problem, and the search of conditions on h that ensure the existence of solutions becomes highly nontrivial. In this section we describe one of these conditions.

Denoting as usual by $\lambda_1 < \lambda_2 \leq \cdots$ the sequence of the eigenvalues of $-\Delta$ in $H_0^1(\Omega)$, we assume that for some n and k,

$$\lambda_n < \lambda_{n+1} = \cdots = \lambda_{n+k} < \lambda_{n+k+1}$$

and we suppose that

$$\lambda = \lambda_{n+1} = \cdots = \lambda_{n+k}. \tag{4.27}$$

So, assumptions (**f$_{10}$**) and (4.27) say that Problem (4.24) is at resonance (with a multiple eigenvalue, if $k \geq 2$).

The hypotheses made so far are not sufficient to prove the existence of a solution. We now describe one type of conditions that guarantee the solvability of (4.24) for certain $h \in L^2(\Omega)$.

These were first introduced in [28], so that they are called the Landesman–Lazer conditions. To state them, let

$$E = \text{span}\{\varphi_{n+1}, \ldots, \varphi_{n+k}\} \tag{4.28}$$

be the k-dimensional subspace of $H_0^1(\Omega)$ made of all the solutions of $-\Delta\varphi = \lambda\varphi$.

The Landesman–Lazer conditions are the following: for every $\varphi \in E \setminus \{0\}$,

$$\textbf{(LL)} \qquad f_r \int_\Omega \varphi^- \, dx - f_l \int_\Omega \varphi^+ \, dx < \int_\Omega h\varphi \, dx < f_l \int_\Omega \varphi^- \, dx - f_r \int_\Omega \varphi^+ \, dx,$$

where of course φ^+ and φ^- denote the positive and negative part of φ.

We notice that since $\varphi \in E$ implies that $-\varphi \in E$, any of the two inequalities in (**LL**) implies the other one.

Remark 4.4.16 To understand the meaning of these inequalities, it is convenient to visualize the case $f_l = -f_r$, with $f_r < 0$. In this case (**LL**) becomes

$$-f_l \int_\Omega \varphi^- \, dx - f_l \int_\Omega \varphi^+ \, dx < \int_\Omega h\varphi \, dx < f_l \int_\Omega \varphi^- \, dx + f_l \int_\Omega \varphi^+ \, dx,$$

and, recalling that $\varphi^+ + \varphi^- = |\varphi|$,

$$-f_l \int_\Omega |\varphi| \, dx < \int_\Omega h\varphi \, dx < f_l \int_\Omega |\varphi| \, dx.$$

This is a sort of generalized orthogonality condition: the (L^2) scalar product between h and φ must not be too large, compared to the limits of f at $\pm\infty$. Note also that if h is orthogonal to E, then the preceding inequality is always satisfied. Compare with Remark 4.4.15.

As in the preceding section, weak solutions of (4.24) are critical points of the functional

$$J(u) = \frac{1}{2}\int_\Omega |\nabla u|^2 - \frac{\lambda}{2}\int_\Omega u^2\, dx - \int_\Omega F(u)\, dx - \int_\Omega hu\, dx,$$

where $F(t) = \int_0^t f(s)\, ds$. Again, due to $(\mathbf{f_{10}})$, the functional J is in $C^{1,1}(H_0^1(\Omega))$.
We are going to prove the following result.

Theorem 4.4.17 *Let $\Omega \subset \mathbb{R}^N$ be open and bounded. Assume that $(\mathbf{f_{10}})$ holds and that the limits f_l and f_r exist and satisfy (4.25). Let λ be as in (4.27). Then for every $h \in L^2(\Omega)$ satisfying (**LL**), Problem (4.24) admits at least one weak solution.*

We are going to prove this theorem by an application of the Saddle Point Theorem.

It is remarkable that the proof of this result as it appeared in the original paper [28] is considerably different and more complicated than the one we give here. This is because at the time of [28] the Saddle Point Theorem was not yet available.

As usual we divide the proof in two problems: the analysis of the compactness properties of J and the verification of the required geometrical assumptions. It is interesting to remark that the resonance assumption affects *both* problems, making proofs more difficult.

We start with the analysis of compactness. Recalling (4.28), we split

$$H_0^1(\Omega) = E \oplus E^\perp$$

and we write $u \in H_0^1(\Omega)$ as $u = w + v$, with $w \in E$ and $v \in E^\perp$.

Lemma 4.4.18 *The functional J satisfies $(PS)_c$ for every $c \in \mathbb{R}$.*

Proof A part of the proof works exactly as that of Lemma 4.4.12, and we will not repeat all the computations.
Let $\{u_k\}$ be a Palais–Smale sequence at level $c \in \mathbb{R}$, namely,

$$J(u_k) \to c \quad \text{and} \quad J'(u_k) \to 0.$$

If $\{u_k\}$ is bounded, then, up to subsequences, it converges strongly in $H_0^1(\Omega)$, exactly as at the end of the proof of Lemma 4.4.12. So the proof reduces to showing that u_k is bounded. Now if we write $u_k = w_k + v_k$, with $w_k \in E$ and $v_k \in E^\perp$, we have of course

$$\int_\Omega \nabla w_k \cdot \nabla z\, dx - \lambda \int_\Omega w_k z\, dx = 0$$

for all $z \in H_0^1(\Omega)$, by definition of λ and E. Therefore the condition $J'(u_k) \to 0$ reads

$$\int_\Omega \nabla v_k \cdot \nabla z \, dx - \lambda \int_\Omega v_k z \, dx - \int_\Omega f(u_k) z \, dx - \int_\Omega h z \, dx = o(1)\|z\| \qquad (4.29)$$

for every $z \in H_0^1(\Omega)$. This is the form that (4.20) takes in the resonance case. Repeating step by step the computations that follow (4.20), one shows that v_k is bounded (if $\|v_k\| \to \infty$, then as in Lemma 4.4.12 one shows that $v_k/\|v_k\|$ converges to an eigenfunction relative to λ in $E^\perp$, which is impossible).

Thus the real problem is to show that w_k is bounded. In the proof of Lemma 4.4.12, this problem is not present because $E = \{0\}$.

So we assume for contradiction that $\|w_k\| \to \infty$. The sequence $w_k/\|w_k\|$ is bounded and, up to subsequences, it converges strongly in $H_0^1(\Omega)$ (because E has finite dimension) and almost everywhere in Ω to some $\varphi \in E$. Since the convergence is strong, $\varphi \not\equiv 0$.

Since v_k is bounded in $H_0^1(\Omega)$, we can assume that up to subsequences, it converges to some v a.e. in Ω. Then we see that

$$u_k(x) = \|w_k\| \frac{w_k(x)}{\|w_k\|} + v_k(x) \to \begin{cases} +\infty & \text{if } \varphi(x) > 0, \\ -\infty & \text{if } \varphi(x) < 0, \end{cases} \quad \text{a.e. in } \Omega. \qquad (4.30)$$

In the preceding relation we have used the fact that φ, being an eigenfunction, is almost everywhere different from zero, as we stated in Theorem 1.7.3. Define a function $f_\infty : \Omega \to \mathbb{R}$ as

$$f_\infty(x) = \begin{cases} f_r & \text{if } \varphi(x) > 0, \\ f_l & \text{if } \varphi(x) < 0. \end{cases} \qquad (4.31)$$

Then from (4.30) we see that

$$f(u_k(x)) \to f_\infty(x) \quad a.e. \text{ in } \Omega,$$

and since f is bounded, by dominated convergence we have

$$f(u_k) \to f_\infty$$

in every $L^p(\Omega)$ with p finite.

Now we choose $z = \varphi$ in (4.29) and we let $k \to \infty$: we obtain

$$-\int_\Omega f_\infty \varphi \, dx - \int_\Omega h\varphi \, dx = 0,$$

namely

$$\int_\Omega h\varphi \, dx = -\int_\Omega f_\infty \varphi \, dx = f_l \int_\Omega \varphi^- \, dx - f_r \int_\Omega \varphi^+ \, dx.$$

This contradicts (**LL**), and shows that also w_k must be bounded. The proof is complete. $\qquad \square$

Now we check that the geometrical assumptions of the Saddle Point Theorem hold.

To this aim we set, as in the nonresonant case,

$$E_n = \text{span}\{\varphi_1, \ldots, \varphi_n\}, \quad \text{and} \quad V = E_n^\perp = \text{cl}\{\text{span}\{\varphi_k \mid k \geq n+1\}\}. \quad (4.32)$$

Notice that the space E used in the previous lemma is now part of V; this is what makes the estimates more complex.

First the easy part. Since the inequality

$$\|u\|^2 \leq \lambda_n |u|_2^2 \quad \forall u \in E_n$$

still holds (see (4.23)), denoting by B_R^n the ball in E_n of center zero and radius R, one can prove exactly as in the nonresonant case, that

$$\max_{u \in \partial B_R^n} J(u) \to -\infty$$

if $R \to \infty$.

Now we turn to the hard part, the estimate from below on V.

Lemma 4.4.19 *We have*

$$\inf_{u \in V} J(u) > -\infty.$$

Proof We write V as

$$V = E \oplus F,$$

where E is the subspace defined in (4.28), and $F = \text{cl}\{\text{span}\{\varphi_j \mid j \geq n+k+1\}\}$. Notice that F is the orthogonal complement of E in V. Accordingly, we write each $u \in V$ as

$$u = w + v,$$

where $w \in E$ and $v \in F$. Recall that w and v are orthogonal *both* in $H_0^1(\Omega)$ and in $L^2(\Omega)$. Also,

$$\|v\|^2 \geq \lambda_{n+k+1} |v|_2^2 \quad \forall v \in F,$$

see (4.23).

Bearing all this in mind, we see that for every $u \in V$,

$$\begin{aligned}
J(u) &= \frac{1}{2} \int_\Omega |\nabla u|^2 - \frac{\lambda}{2} \int_\Omega u^2 \, dx - \int_\Omega F(u) \, dx - \int_\Omega hu \, dx \\
&\geq \frac{1}{2}\left(1 - \frac{\lambda}{\lambda_{n+k+1}}\right) \int_\Omega |\nabla v|^2 \, dx - \int_\Omega F(u) \, dx - \int_\Omega hu \, dx \\
&= \mu \|v\|^2 - \int_\Omega F(u) \, dx - \int_\Omega hu \, dx, \quad (4.33)
\end{aligned}$$

with $\mu > 0$.

Since f is bounded, $|F(t)| \leq b|t|$ for every $t \in \mathbb{R}$ and for some constant b. Therefore from the preceding inequality we obtain

$$J(u) \geq \mu \|v\|^2 - C\|u\| - |h|_2 |u|_2 \geq \mu \|v\|^2 - C\|v\| - C\|w\|, \quad (4.34)$$

for some positive constant C independent of u.

Assume for contradiction that there exists a sequence $u_k = w_k + v_k \in V$ such that

$$J(u_k) \to -\infty \tag{4.35}$$

as $k \to \infty$. Then we claim that

$$\frac{\|w_k\|}{\|v_k\|} \to +\infty. \tag{4.36}$$

We assume here that $\|v_k\| \neq 0$; if this is not the case the proof is simpler and we omit the details. To check (4.36), notice that from (4.34) and (4.35), we have

$$\mu\|v_k\|^2 - C\|v_k\| - C\|w_k\| \to -\infty,$$

so that necessarily $\|w_k\| \to \infty$. If $\|v_k\|$ is bounded, we are done. If $\|v_k\| \to \infty$, then writing the preceding relation as

$$\|v_k\|\left(\mu\|v_k\| - C - C\frac{\|w_k\|}{\|v_k\|}\right) \to -\infty$$

shows that (4.36) must be true.

Now, as in the preceding lemma, we can assume that there exist $\varphi \in E \setminus \{0\}$ and $v \in F$ such that (up to subsequences),

$$\frac{w_k(x)}{\|w_k\|} \to \varphi(x) \quad \text{and} \quad \frac{v_k(x)}{\|v_k\|} \to v(x)$$

almost everywhere in Ω. Note that $\varphi(x) \not\equiv 0$, since the convergence of $w_k/\|w_k\|$ is strong, as above. Therefore, by (4.36),

$$u_k(x) = \|w_k\|\left(\frac{w_k(x)}{\|w_k\|} + \frac{\|v_k\|}{\|w_k\|}\frac{v_k(x)}{\|v_k\|}\right)$$

$$\to \begin{cases} +\infty & \text{if } \varphi(x) > 0, \\ -\infty & \text{if } \varphi(x) < 0, \end{cases} \quad \text{a.e. in } \Omega. \tag{4.37}$$

Next we claim that

$$\frac{1}{\|w_k\|}F(u_k(x)) \to f_\infty\varphi(x) \quad \text{a.e. in } \Omega, \tag{4.38}$$

where f_∞ is the function defined in (4.31). To see this note first that

$$\lim_{t \to -\infty}\frac{F(t)}{t} = f_l \quad \text{and} \quad \lim_{t \to +\infty}\frac{F(t)}{t} = f_r,$$

by the de l'Hôpital Theorem. Now for almost every x in the set where $\varphi(x) > 0$, we have, for all k large,

$$\frac{1}{\|w_k\|}F(u_k(x)) = \frac{u_k(x)}{\|w_k\|}\frac{F(u_k(x))}{u_k(x)} = \left(\frac{v_k(x)}{\|w_k\|} + \frac{w_k(x)}{\|w_k\|}\right)\frac{F(u_k(x))}{u_k(x)}.$$

When $k \to \infty$, from the preceding discussions we see that

$$\frac{v_k(x)}{\|w_k\|} = \frac{\|v_k\|}{\|w_k\|}\frac{v_k(x)}{\|v_k\|} \to 0, \qquad \frac{w_k(x)}{\|w_k\|} \to \varphi(x), \qquad \frac{F(u_k(x))}{u_k(x)} \to f_r$$

almost everywhere where $\varphi(x) > 0$. The last limit of course is f_l a.e. in the set where $\varphi(x) < 0$. Then (4.38) is proved.

Now we check that the convergence in (4.38) actually holds in $L^1(\Omega)$.

The function $u_k/\|u_k\|$ is bounded in $H_0^1(\Omega)$, and therefore, up to subsequences, it converges strongly in $L^1(\Omega)$, and there exists $u \in L^1(\Omega)$ such that

$$\frac{|u_k(x)|}{\|u_k\|} \leq u(x) \quad \text{a.e. in } \Omega.$$

Then

$$\frac{|F(u_k(x))|}{\|w_k\|} \leq b\frac{|u_k(x)|}{\|w_k\|} = b\frac{\|u_k\|}{\|w_k\|}\frac{|u_k(x)|}{\|u_k\|} \leq C + Cu(x) \in L^1(\Omega)$$

a.e. in Ω, because by (4.36) the coefficient $\|u_k\|/\|w_k\|$ is bounded independently of k. Then by dominated convergence,

$$\frac{F(u_k)}{\|w_k\|} \to f_\infty\varphi \quad \text{in } L^1(\Omega).$$

We are now ready to conclude. Since $u_k/\|w_k\|$ is bounded, up to subsequences it converges weakly in $H_0^1(\Omega)$ and strongly in $L^2(\Omega)$; by (4.37), it must be

$$\frac{u_k}{\|w_k\|} \to \varphi.$$

Then from (4.33) we see that

$$J(u_k) \geq \mu\|v_k\|^2 - \int_\Omega F(u_k)\,dx - \int_\Omega hu_k\,dx$$

$$= \mu\|v_k\|^2 - \|w_k\|\left(\frac{1}{\|w_k\|}\int_\Omega F(u_k)\,dx + \int_\Omega h\frac{u_k}{\|w_k\|}\,dx\right)$$

$$\geq -\|w_k\|\left(\int_\Omega f_\infty\varphi\,dx + \int_\Omega h\varphi\,dx + o(1)\right)$$

$$= -\|w_k\|\left(f_r\int_\Omega \varphi^+\,dx - f_l\int_\Omega \varphi^-\,dx + \int_\Omega h\varphi\,dx + o(1)\right).$$

By **(LL)**, the quantity in the round bracket is strictly negative for large k, so that we obtain

$$J(u_k) \to +\infty$$

as $k \to \infty$, contradicting (4.35). The proof is complete. $\square$

End of the proof of Theorem 4.4.17 The functional J satisfies the Palais–Smale condition at every level and the geometrical assumptions of the Saddle Point Theorem if we choose E_n and V as in (4.32). Then J has a critical point, namely Problem (4.24) has a weak solution. $\square$

Remark 4.4.20 The Landesman–Lazer conditions **(LL)** may seem a bit mysterious at first glance. However they arise naturally as *necessary* conditions for the existence of a solution to (4.24), under a slightly stronger assumption. Indeed, if on f we also assume that

$$f_r < f(t) < f_l \quad \forall t \in \mathbb{R}, \tag{4.39}$$

then we can prove that the existence of a solution to (4.24) *implies* (**LL**). To see this, let u be a solution of Problem (4.24). Then for every $\varphi \in E$,

$$0 = \int_\Omega \nabla u \cdot \nabla \varphi \, dx - \lambda \int_\Omega u\varphi \, dx = \int_\Omega f(u)\varphi \, dx + \int_\Omega h\varphi \, dx,$$

so that

$$\int_\Omega h\varphi \, dx = -\int_\Omega f(u)\varphi \, dx = \int_\Omega f(u)\varphi^- \, dx - \int_\Omega f(u)\varphi^+ \, dx.$$

By (4.39),

$$\int_\Omega h\varphi \, dx = \int_\Omega f(u)\varphi^- \, dx - \int_\Omega f(u)\varphi^+ \, dx < f_l \int_\Omega \varphi^- \, dx - f_r \int_\Omega \varphi^+ \, dx$$

and, similarly,

$$\int_\Omega h\varphi \, dx = \int_\Omega f(u)\varphi^- \, dx - \int_\Omega f(u)\varphi^+ \, dx > f_r \int_\Omega \varphi^- \, dx - f_l \int_\Omega \varphi^+ \, dx,$$

and these are the Landesman–Lazer conditions.

4.5 Problems with a Parameter

In many mathematical problems deriving from applications the presence of one parameter (or more) is a relevant feature, and the study of how solutions depend on parameters is an important topic. Most of the work on such questions uses tools different from those explained in this book, such as bifurcation theory; we refer for example to [2, 4] for a more extensive treatment of this matters. However some interesting results can be obtained also in a variational framework, and in this section we give some examples.

We start with a very simple model case where an insight on the problem is obtained by elementary computations. Consider the problem

$$\begin{cases} -\Delta u + q(x)u = \lambda |u|^{p-2}u & \text{in } \Omega, \\ u = 0 & \text{on } \partial\Omega, \end{cases} \tag{4.40}$$

with the usual hypotheses: Ω is a bounded open set, $q \in L^\infty(\Omega)$ and $q \geq 0$, $p \in (2, 2^*)$. We know, by the theorems of Sects. 2.3 or 4.3, that (4.40) admits a nonnegative and nontrivial solution for all $\lambda > 0$. Let u be a solution for $\lambda = 1$. It is then immediate to see that $u_\lambda = \lambda^{-\frac{1}{p-2}}u$ is a solution of (4.40). Hence we have a family $\{u_\lambda\}_\lambda$ of solutions such that $\|u_\lambda\| \to 0$ when $\lambda \to +\infty$ and $\|u_\lambda\| \to +\infty$ when $\lambda \to 0^+$. We now want to show that a similar result holds in a more general framework. So we consider the *nonlinear eigenvalue problem*

$$\begin{cases} -\Delta u + q(x)u = \lambda f(u) & \text{in } \Omega, \\ u = 0 & \text{on } \partial\Omega. \end{cases} \tag{4.41}$$

Denoting as usual $F(t) = \int_0^t f(s)\,ds$, we can prove the following result. In its statement, and in all the statements of the next theorems, the positivity of q in assumption $(\mathbf{h_1})$ can of course be replaced by (2.3).

Theorem 4.5.1 *Assume that* $(\mathbf{h_1})$, $(\mathbf{f_3})$, $(\mathbf{f_6})$ *and* $(\mathbf{f_7})$ *hold. Then the following statements hold.*

(i) *If there exist $r > 2$, $M_1 > 0$ and $t_1 > 0$ such that $f(t) \geq M_1 t^{r-1}$ for $t \in [0, t_1]$, then for every $\lambda > 0$ there exists a nonnegative and non trivial solution u_λ of (4.41) such that $\|u_\lambda\| \to 0$ as $\lambda \to +\infty$.*

(ii) *If $F(t) > 0$ for all $t > 0$ then for every $\lambda > 0$ there exists a nonnegative and non trivial solution u_λ of (4.41) such that $\|u_\lambda\| \to +\infty$ as $\lambda \to 0^+$.*

Proof As we are interested in nonnegative solutions, we can assume that $f(t) = 0$ for all $t \leq 0$ (recall Example 1.7.10). The existence result is a consequence of the Mountain Pass Theorem 4.3.1 applied to the functional

$$J_\lambda(u) = \frac{1}{2}\|u\|^2 - \lambda \int_\Omega F(u)\,dx.$$

Arguing as we have done in the proof of Theorem 4.4.3, it is easy to see that J_λ is a $C^{1,1}$ functional over $H_0^1(\Omega)$ satisfying *(PS)*. Moreover, the assumptions of the theorem imply $(\mathbf{f_6})$–$(\mathbf{f_9})$, so that we can apply Theorem 4.4.3 and obtain a nontrivial and nonnegative solution. All we have to do is to show that we can choose a mountain pass solution satisfying the requested asymptotic properties. This will be done by constructing some particular minimax classes separately for statements (i) and (ii).

(i) We study the case $\lambda \to +\infty$. Without loss of generality we can assume that $F(t) \geq M_1 t^r$ for all $t \in [0, t_1]$. Also recall that $F(t) = 0$ for all $t \leq 0$. Now fix a function $\psi \in H_0^1(\Omega) \backslash \{0\}$ such that $0 \leq \psi(x) \leq t_1$ for all $x \in \Omega$. We have

$$J_\lambda(\psi) \leq \frac{1}{2}\|\psi\|^2 - \lambda M_1 \int_\Omega \psi^r\,dx,$$

so that there exists $\bar{\lambda} > 0$ such that $J_\lambda(\psi) < 0$ for all $\lambda \geq \bar{\lambda}$. Let us now investigate what happens close to zero. Arguing as in the proof of Theorem 4.4.3, we see that

$$J_\lambda(u) \geq \frac{1}{4}\|u\|^2 - C\lambda\|u\|^{2^*} = \|u\|^2\left(\frac{1}{4} - C\lambda\|u\|^{2^*-2}\right),$$

where C does not depend on u nor on λ. We define

$$\rho_\lambda = \left(\frac{1}{8C\lambda}\right)^{\frac{1}{2^*-2}}$$

and we notice that, for large λ, we certainly have $\rho_\lambda < \|\psi\|$. Clearly, when $\|u\| = \rho_\lambda$,

$$J_\lambda(u) \geq \frac{\rho_\lambda^2}{8} > 0.$$

Setting $\alpha_\lambda = \frac{\rho_\lambda^2}{8}$, we have obtained some particular values for which the geometrical hypotheses of the Mountain Pass Theorem are satisfied. So for large λ we can define

$$\Gamma = \{\gamma([0,1]) \mid \gamma : [0,1] \to H_0^1(\Omega) \text{ continuous, } \gamma(0) = 0, \gamma(1) = \psi\}$$

and

$$c_\lambda = \inf_{\gamma \in \Gamma} \max_{t \in [0,1]} J_\lambda(\gamma(t)),$$

and we obtain that $c_\lambda > 0$ is a critical value for J_λ. Let u_λ be a critical point of J_λ at level c_λ (that is, a solution of (4.41)) such that $J_\lambda(u_\lambda) = c_\lambda$. We claim that $c_\lambda \to 0$ as $\lambda \to +\infty$. To see this, we first notice that

$$c_\lambda \leq \max_{t \in [0,1]} J_\lambda(t\psi),$$

and, for $t \in [0,1]$,

$$J_\lambda(t\psi) = t^2 \frac{1}{2} \|\psi\|^2 - \lambda \int_\Omega F(t\psi)\,dx \leq t^2 \frac{1}{2}\|\psi\|^2 - \lambda M_1 t^r \int_\Omega \psi^r\,dx.$$

Consider the real function

$$\eta(t) = t^2 \frac{1}{2} \|\psi\|^2 - \lambda M_1 t^r \int_\Omega \psi^r\,dx,$$

on the interval $[0,1]$. By elementary computations we see that η attains its maximum at the point

$$t_\lambda = \left(\frac{\|\psi\|^2}{\lambda r M_1 |\psi|_r^r} \right)^{\frac{1}{r-2}},$$

and that the value of this maximum is

$$\eta(t_\lambda) = t_\lambda^2 \frac{1}{2} \|\psi\|^2 - \lambda M_1 t_\lambda^r \int_\Omega \psi^r\,dx = a \frac{1}{\lambda^{\frac{2}{r-2}}} - b \frac{1}{\lambda^{\frac{2}{r-2}}},$$

where a, b are positive constants depending on ψ. This implies that

$$\max_{t \in [0,1]} J_\lambda(t\psi) \leq \max_{t \in [0,1]} \eta(t) = (a - b) \frac{1}{\lambda^{\frac{2}{r-2}}},$$

and hence

$$0 < c_\lambda \leq (a - b) \frac{1}{\lambda^{\frac{2}{r-2}}},$$

so that $c_\lambda \to 0$ as $\lambda \to \infty$, as we claimed. From this it is easy to deduce that $\|u_\lambda\| \to 0$. Indeed, we have

$$c_\lambda = J_\lambda(u_\lambda) = \frac{1}{2} \|u_\lambda\|^2 - \lambda \int_\Omega F(u_\lambda)\,dx$$

and

$$0 = \frac{1}{\mu} J_\lambda'(u_\lambda)u_\lambda = \frac{1}{\mu} \|u_\lambda\|^2 - \frac{\lambda}{\mu} \int_\Omega f(u_\lambda)u_\lambda.$$

Subtracting and using $(\mathbf{f_3})$, we see that

$$c_\lambda = \left(\frac{1}{2} - \frac{1}{\mu}\right)\|u_\lambda\|^2 + \lambda \int_\Omega \left(\frac{1}{\mu} f(u_\lambda)u_\lambda - F(u_\lambda)\right) dx \geq \left(\frac{1}{2} - \frac{1}{\mu}\right)\|u_\lambda\|^2.$$

This shows that $\|u_\lambda\| \to 0$ as $\lambda \to +\infty$.

(ii) We now study the case $\lambda \to 0^+$. From $(\mathbf{f_6})$ and $(\mathbf{f_7})$ we deduce that there are constants M_2, M_3 such that

$$F(t) \leq M_3 t^2 + M_2 t^p$$

for all $t \geq 0$. Also, from $(\mathbf{f_3})$ and that fact that $F(t) > 0$, we see that $F(t) \geq Ct^\mu$ for all $t > 0$, and of course $\mu \leq p$. Let us fix $\psi \in H_0^1(\Omega)\setminus\{0\}$ such that $\psi(x) \geq 0$ for all $x \in \Omega$. We have, for every $t \geq 0$,

$$J_\lambda(t\psi) \leq t^2 \frac{1}{2}\|\psi\|^2 - \lambda Ct^\mu \int_\Omega \psi^\mu \, dx.$$

For every fixed λ we want to choose t_λ such that $J_\lambda(t_\lambda\psi) < 0$. From the above computations we easily see that we can choose any t such that

$$t > \left(\frac{\|\psi\|^2}{2\lambda C \int_\Omega \psi^\mu \, dx}\right)^{\frac{1}{\mu-2}} = \left(\frac{1}{\lambda}\right)^{\frac{1}{\mu-2}} a_1,$$

where $a_1 > 0$ is a constant depending on ψ. To fix ideas, we choose

$$t_\lambda = 2\left(\frac{1}{\lambda}\right)^{\frac{1}{\mu-2}} a_1.$$

Then we have

$$J_\lambda(t_\lambda\psi) < 0 \quad \text{and} \quad \|t_\lambda\psi\| = \frac{1}{\lambda^{\frac{1}{\mu-2}}} a_2,$$

where a_2 does not depend on λ, and of course $\|t_\lambda\psi\| \to +\infty$ as $\lambda \to 0^+$. As in case (i) we estimate

$$J_\lambda(u) \geq \frac{1}{4}\|u\|^2 - C\lambda\|u\|^{2^*} = \|u\|^2\left(\frac{1}{4} - C\lambda\|u\|^{2^*-2}\right),$$

so $J_\lambda(u) \geq \frac{1}{8}$ for small λ and $\|u\| = 1$. If we choose $\rho = 1$ and $\alpha = \frac{1}{8}$, for small λ we obtain $\|t_\lambda\psi\| > \rho$ and we have again some particular values for which the geometrical hypotheses of Mountain Pass Theorem are satisfied. So we define, for every $\lambda > 0$,

$$\Gamma_\lambda = \{\gamma([0,1]) \mid \gamma : [0,1] \to H_0^1(\Omega) \text{ continuous}, \gamma(0) = 0, \gamma(1) = t_\lambda\psi\}$$

and

$$c_\lambda = \inf_{\gamma \in \Gamma_\lambda} \max_{t \in [0,1]} J_\lambda(\gamma(t)),$$

and we obtain a critical point u_λ of J_λ at level c_λ. We claim now that $c_\lambda \to +\infty$ as $\lambda \to 0^+$. To see this, let $\varepsilon \in (0,1)$ to be fixed later. By continuity, for every $\gamma \in \Gamma_\lambda$ there exists $\tilde{t} \in [0,1]$ such that

$$\|\gamma(\tilde{t})\| = \varepsilon\|t_\lambda\psi\| = \frac{\varepsilon}{\lambda^{\frac{1}{\mu-2}}} a_2.$$

Setting $v = \gamma(\tilde{t})$ we estimate

$$
\begin{aligned}
J_\lambda(\gamma(\tilde{t})) &= \frac{1}{2}\|v\|^2 - \lambda \int_\Omega F(v)\,dx \\
&\geq \frac{\varepsilon^2}{\lambda^{\frac{2}{\mu-2}}}a_3 - \lambda M_3 \int_\Omega v^2\,dx - \lambda M_2 \int_\Omega |v|^p\,dx \\
&\geq \frac{\varepsilon^2}{\lambda^{\frac{2}{\mu-2}}}a_3 - \lambda C\|v\|^2 - \lambda C\|v\|^p \\
&= \frac{\varepsilon^2}{\lambda^{\frac{2}{\mu-2}}}a_3 - \lambda\frac{\varepsilon^2}{\lambda^{\frac{2}{\mu-2}}}a_4 - \lambda\frac{\varepsilon^p}{\lambda^{\frac{p}{\mu-2}}}a_5,
\end{aligned}
$$

where a_i are positive and independent of λ, $i = 3, 4, 5$. Noticing that $0 \leq \frac{p-\mu}{p-2} < 1$, we take

$$
\alpha \in \left(\frac{p-\mu}{p-2}, 1\right)
$$

and for every $0 < \lambda < 1$ we choose a particular value for ε, namely

$$
\varepsilon = \lambda^{\frac{\alpha}{\mu-2}}.
$$

By elementary computations we obtain

$$
J_\lambda(\gamma(\tilde{t})) \geq \frac{1}{\lambda^{\frac{2-2\alpha}{\mu-2}}}a_3 - \frac{1}{\lambda^{\frac{2-2\alpha}{\mu-2}-1}}a_4 - \frac{1}{\lambda^{\frac{p-p\alpha}{\mu-2}-1}}a_5 =: b(\lambda),
$$

so that

$$
\max_{t\in[0,1]} J_\lambda(\gamma(t)) \geq b(\lambda).
$$

Since this holds for every $\gamma \in \Gamma_\lambda$, we deduce that

$$
c_\lambda \geq b(\lambda).
$$

On the other hand, as $\alpha < 1$, we have

$$
\frac{2-2\alpha}{\mu-2} > 0,
$$

while $\alpha > \frac{p-\mu}{p-2}$ implies

$$
\frac{2-2\alpha}{\mu-2} > \frac{p-p\alpha}{\mu-2} - 1.
$$

Therefore $b(\lambda) \to +\infty$ as $\lambda \to 0^+$, which implies that $c_\lambda \to +\infty$ as $\lambda \to 0^+$, and the claim is proved.

To conclude the proof of part (ii), let $\{\lambda_k\}_k$ be a sequence such that $\lambda_k \to 0^+$ as $k \to +\infty$. Assume for contradiction that $\{\|u_{\lambda_k}\|\}_k$ is bounded. It is then easy to see that also $J_{\lambda_k}(u_{\lambda_k})$ is bounded, a contradiction. We deduce that for every sequence $\{\lambda_k\}_k$ such that $\lambda_k \to 0^+$, the sequence $\{\|u_{\lambda_k}\|\}_k$ is unbounded. This means that $\|u_\lambda\| \to +\infty$ as $\lambda \to 0^+$. $\square$

Remark 4.5.2 If $2 < p < r < 2^*$ the function f defined by $f(t) = t^{p-1} + t^{r-1}$ for $t > 0$ and $f(t) = 0$ for $t \le 0$ satisfies all the hypotheses of the preceding theorem. For such a nonlinearity we then have a family $\{u_\lambda\}$ of non negative mountain pass solutions of (4.41) such that $\|u_\lambda\| \to 0$ as $\lambda \to +\infty$ and $\|u_\lambda\| \to +\infty$ as $\lambda \to 0^+$. This f is not homogeneous, so the simple scaling argument that we used at the beginning of the section for the nonlinearity $f(t) = t^{p-1}$ does not work.

We now consider Problem (4.41) with a different nonlinearity f, exhibiting a different power-like behavior near zero and near infinity. We consider $1 < r < 2 < s$ and we define a function $f : \mathbb{R} \to \mathbb{R}$ as

$$f(t) = \begin{cases} t^{s-1} & \text{if } t \in [0, 1], \\ t^{r-1} & \text{if } t \ge 1, \\ 0 & \text{if } t \le 0. \end{cases} \tag{4.42}$$

As usual we set $F(t) = \int_0^t f(s)\,ds$, so that

$$F(t) = \begin{cases} \frac{1}{s}t^s & \text{if } t \in [0, 1], \\ \frac{1}{r}t^r - \frac{1}{r} + \frac{1}{s} & \text{if } t \ge 1, \\ 0 & \text{if } t \le 0. \end{cases} \tag{4.43}$$

Notice that $0 \le f(t) \le |t|$ for all t, so the usual arguments show that the functional

$$u \mapsto \int_\Omega F(u)\,dx$$

is well defined in $H_0^1(\Omega)$ and its differential is locally Lipschitz continuous. We want to study existence, multiplicity and asymptotic behavior of nonnegative non trivial solutions of (4.41), for $\lambda > 0$. We will look for critical points of the $C^{1,1}$ functional

$$J_\lambda(u) = \frac{1}{2}\|u\|^2 - \lambda \int_\Omega F(u)\,dx. \tag{4.44}$$

The functional J_λ is coercive (see below the proof of Theorem 4.5.4), and from this it is easy to see that it satisfies the *(PS)* condition (see Exercise 10).

First we give a "negative result".

Proposition 4.5.3 *Assume that* ($\mathbf{h_1}$) *holds. Then for all* $\lambda \in [0, \lambda_1]$ *the only nonnegative solution of* (4.41) *is the trivial one.*

Proof If $J'_\lambda(u) = 0$ and u is nonnegative, then

$$\|u\|^2 = \lambda \int_\Omega f(u)u\,dx \le \lambda \int_\Omega u^2\,dx \le \frac{\lambda}{\lambda_1}\|u\|^2$$

by the Poincaré inequality. Therefore $\lambda \ge \lambda_1$. The case $\lambda = \lambda_1$ is impossible because the preceding inequality would become an equality, and so $u \equiv t\varphi_1$ for some $t \in \mathbb{R}$. But $t\varphi_1$ solves the equation only if $t = 0$. Therefore the existence of a nontrivial nonnegative solution implies $\lambda > \lambda_1$.   □

Now we show that a first nontrivial solution exists provided λ is large enough.

Theorem 4.5.4 *Assume that* (**h₁**) *holds. Then there exists* $\lambda_* > 0$ *such that for every* $\lambda > \lambda_*$ *there exists a nonnegative solution* v_λ *to* (4.41) *such that*

$$J_\lambda(v_\lambda) = \min_{u \in H_0^1(\Omega)} J_\lambda(u) < 0.$$

Proof Choose any $v \in H_0^1(\Omega) \backslash \{0\}$ such that $v \geq 0$. It is then obvious that

$$J_\lambda(v) \to -\infty \quad \text{as } \lambda \to +\infty,$$

so that also

$$\inf_{u \in H_0^1(\Omega)} J_\lambda(u) \to -\infty \quad \text{as } \lambda \to +\infty.$$

Hence we can fix $\lambda_* > 0$ such that $\inf_{u \in H_0^1(\Omega)} J_\lambda(u) < 0$ for all $\lambda > \lambda_*$.

On the other hand, for every fixed λ, J_λ is bounded from below. Indeed, for all $u \in H_0^1(\Omega)$ we have

$$
\begin{aligned}
J_\lambda(u) &= \frac{1}{2}\|u\|^2 - \lambda \int_\Omega F(u)\,dx \\
&= \frac{1}{2}\|u\|^2 - \frac{\lambda}{s}\int_{\{0 \leq u \leq 1\}} u^s\,dx - \frac{\lambda}{r}\int_{\{1 \leq u\}} u^r\,dx - \lambda\left(\frac{1}{s} - \frac{1}{r}\right)|\{u \geq 1\}| \\
&\geq \frac{1}{2}\|u\|^2 - \lambda C - \frac{\lambda}{r}\int_\Omega |u|^r\,dx \geq \frac{1}{2}\|u\|^2 - \lambda C - \lambda C\|u\|^r,
\end{aligned}
$$

which implies that J_λ is coercive because $r < 2$. By the usual arguments of Chap. 2, we obtain that J_λ has a minimum value which is attained by some v_λ.

For $\lambda > \lambda_*$ of course

$$J_\lambda(v_\lambda) = \inf_{u \in H_0^1(\Omega)} J_\lambda(u) < 0,$$

and so v_λ is not identically zero. Since v_λ is a solution of (4.41), then $v_\lambda \geq 0$, because $f(t) = 0$ for $t \leq 0$ (see Proposition 1.7.9). □

To find a *second* nontrivial solution we show that J_λ satisfies the assumption of the Mountain Pass Theorem near zero. Precisely, we fix a function $\psi \in H_0^1(\Omega)\backslash\{0\}$ such that $0 \leq \psi \leq 1$. For large λ's we have $J_\lambda(\psi) < 0$. We now show that we can find a second solution by a mountain pass procedure on the paths joining 0 to ψ.

Lemma 4.5.5 *For large* λ's *there exist* $\alpha_\lambda, \rho_\lambda > 0$ *such that*

$$\|u\| = \rho_\lambda \quad \text{implies} \quad J_\lambda(u) \geq \alpha_\lambda.$$

Moreover $\|\psi\| > \rho_\lambda$.

Proof Setting $p = \min\{s, 2^*\}$ we have $2 < p \leq 2^*$ and $0 \leq f(t) \leq |t|^{p-1}$ for all t, so that

$$J_\lambda(u) = \frac{1}{2}\|u\|^2 - \frac{\lambda}{s}\int_{\{0 \le u \le 1\}} u^s\, dx - \frac{\lambda}{r}\int_{\{1 \le u\}} u^r\, dx - \lambda\left(\frac{1}{s} - \frac{1}{r}\right)\big|\{u \ge 1\}\big|$$

$$\ge \frac{1}{2}\|u\|^2 - \frac{\lambda}{s}\int_{\{0 \le u \le 1\}} u^p\, dx - \frac{\lambda}{r}\int_{\{1 \le u\}} u^p\, dx$$

$$\ge \frac{1}{2}\|u\|^2 - \lambda\left(\frac{1}{s} + \frac{1}{r}\right)\int_\Omega |u|^p\, dx \ge \frac{1}{2}\|u\|^2 - \lambda C\|u\|^p$$

$$= \|u\|^2\left(\frac{1}{2} - \lambda C\|u\|^{p-2}\right).$$

If we choose

$$\rho_\lambda = \left(\frac{1}{4\lambda C}\right)^{\frac{1}{p-2}} \quad \text{and} \quad \alpha_\lambda = \frac{1}{4}\rho_\lambda^2,$$

then we obtain that $J_\lambda(u) \ge \alpha_\lambda > 0$ when $\|u\| = \rho_\lambda$, and $\|\psi\| > \rho_\lambda$ for large λ's. $\square$

Theorem 4.5.6 *Assume that* (**h₁**) *holds. Then for large* λ*'s the functional* J_λ *has two non trivial non negative critical points, a minimum* v_λ *and a mountain pass critical point* u_λ.

Proof Thanks to the preceding lemma it is easy to prove that J_λ satisfies all the hypotheses of the Mountain Pass Theorem. The critical point corresponding to the global minimum is given by Theorem 4.5.4. The two solutions are different because

$$J_\lambda(v_\lambda) < 0 < \alpha_\lambda \le J_\lambda(u_\lambda). \qquad \square$$

Theorem 4.5.7 *As* $\lambda \to +\infty$ *there results* $\|v_\lambda\| \to +\infty$ *and* $\|u_\lambda\| \to 0$.

Proof We first consider the minimum point v_λ. Assume for contradiction that the claim is not true. Then there would be a sequence $\lambda_k \to \infty$ such that $\|v_{\lambda_k}\|$ is bounded. Let us write $v_k = v_{\lambda_k}$. We already know that

$$J_{\lambda_k}(v_k) \to -\infty \quad \text{as } k \to \infty. \tag{4.45}$$

On the other hand we have

$$\|v_k\|^2 = \lambda_k \int_\Omega f(v_k)v_k\, dx = \lambda_k \int_{\{0 \le v_k \le 1\}} v_k^s\, dx + \lambda_k \int_{\{1 \le v_k\}} v_k^r\, dx.$$

By assumption, $\|v_k\|$ is bounded and therefore, if we define

$$a_k = \lambda_k \int_{\{0 \le v_k \le 1\}} v_k^s\, dx \quad \text{and} \quad b_k = \lambda_k \int_{\{1 \le v_k\}} v_k^r\, dx,$$

$\{a_k\}_k$ and $\{b_k\}_k$ are bounded sequences too. If we set

$$c_k = \lambda_k\big|\{v_k \ge 1\}\big|,$$

we obtain $0 \le c_k \le b_k$, which shows that also $\{c_k\}_k$ is bounded. We can then compute

$$J_{\lambda_k}(v_k) = \frac{1}{2}\|v_k\|^2 - \lambda_k \int_\Omega F(v_k)\, dx$$

$$= \frac{1}{2}\|v_k\|^2 - \frac{\lambda_k}{s} \int_{\{0 \le v_k \le 1\}} v_k^s\, dx$$

$$- \frac{\lambda_k}{r} \int_{\{1 \le v_k\}} v_k^r\, dx - \lambda_k \left(\frac{1}{s} - \frac{1}{r}\right)|\{v_k \ge 1\}|$$

$$= \frac{1}{2}\|v_k\|^2 - \frac{1}{s} a_k - \frac{1}{r} b_k - \left(\frac{1}{s} - \frac{1}{r}\right) c_k.$$

This gives that $J_{\lambda_k}(v_k)$ is bounded, contradicting (4.45); thus $\|v_\lambda\|$ cannot be bounded as $\lambda \to 0$.

As far as $\|u_\lambda\|$ is concerned, we recall that this solution is obtained by a minimax procedure on the paths joining $u = 0$ with a fixed function $\psi \in H_0^1(\Omega)\setminus\{0\}$ such that $0 \le \psi \le 1$. Hence, if c_λ is the mountain pass critical level,

$$0 < c_\lambda \le \max_{t \in [0,1]} J_\lambda(t\psi).$$

For $t \in [0, 1]$ we have

$$J_\lambda(t\psi) = \frac{t^2}{2}\|\psi\|^2 - \lambda \frac{t^s}{s} \int_\Omega \psi^s\, dx,$$

and the argument is now exactly the same as that of Theorem 4.5.1.   □

4.6 Exercises

1. Let H be a Hilbert space and let $J \in C^1(H)$. Prove that if J satisfies $(PS)_c$, then the set of critical points at level c, namely

   $$\mathcal{K}_c = \{u \in H \mid J(u) = c,\, J'(u) = 0\}$$

   is compact.
2. Let H be a Hilbert space and let $K : H \to \mathbb{R}$ be C^1 and such that $\nabla K : H \to H$ sends bounded sets into precompact sets. Consider the functional $J : H \to \mathbb{R}$ defined by

   $$J(u) = \frac{1}{2}\|u\|^2 + K(u).$$

   Show that J satisfies the Palais–Smale condition if and only if every Palais–Smale sequence is bounded.
3. Let $p \in (2, 2^*)$ and consider, on $H^1(\mathbb{R}^N)$, the functional

   $$I(u) = \frac{1}{2} \int_{\mathbb{R}^N} |\nabla u|^2\, dx + \frac{1}{2} \int_{\mathbb{R}^N} u^2\, dx - \frac{1}{p} \int_{\mathbb{R}^N} |u|^p\, dx.$$

   (a) Show that u_k is a sequence that tends to 0 (strongly in $H^1(\mathbb{R}^N)$) if and only if u_k is a Palais–Smale sequence at level 0.

(b) Show that the Palais–Smale condition does not hold at any level $c \neq 0$.

4. Let B be the unit ball centered at zero in $\mathbb{R}^N$, with $N \geq 3$, and let $H^1_{0,rad}$ be the space of radial functions in $H^1_0(B)$, with norm $\|u\|^2 = \int_B |\nabla u|^2 \, dx$. Let $\alpha > 0$.

(a) Use Lemma 3.1.2 to show that there exists $C > 0$ such that

$$\int_B |u|^p |x|^\alpha \, dx \leq C \|u\|^p \quad \forall p \in \left(2, 2^* + \frac{2\alpha}{N-2}\right), \forall u \in H^1_{0,rad}.$$

(b) Prove that if u_k tends to u weakly in $H^1_{0,rad}$, then, for every exponent $p \in (2, 2^* + \frac{2\alpha}{N-2})$,

$$\int_B |u_k - u|^p |x|^\alpha \, dx \to 0.$$

Hint: take $q \in (p, 2^* + \frac{2\alpha}{N-2})$, so that $p = 2\theta + (1-\theta)q$ for some $\theta \in (0,1)$ and write

$$|u_k - u|^p |x|^\alpha = |u_k - u|^{2\theta} |x|^{\theta \alpha} \, |u_k - u|^{(1-\theta)q} |x|^{(1-\theta)\alpha}.$$

Use the Hölder inequality and the Sobolev embeddings to conclude.

5. Using the notation and the results of the previous exercise, prove that the *Hénon equation*

$$\begin{cases} -\Delta u = |x|^\alpha |u|^{p-2} u & \text{in } B, \\ u = 0 & \text{on } \partial B, \end{cases}$$

has at least one solution in $H^1_{0,rad}$ for every $p \in (2, 2^* + \frac{2\alpha}{N-2})$ and every $\alpha > 0$ by

(i) applying the Mountain Pass Theorem to the functional

$$I(u) = \frac{1}{2} \int_B |\nabla u|^2 \, dx - \frac{1}{p} \int_B |u|^p |x|^\alpha \, dx$$

(ii) minimizing the functional

$$Q(u) = \frac{\int_B |\nabla u|^2 \, dx}{(\int_B |u|^p |x|^\alpha \, dx)^{2/p}}.$$

This problem has been solved in [37].

6. Assume that $\Omega \subset \mathbb{R}^N$ is open and bounded, and let $f : [0, +\infty) \to [0, +\infty)$ be continuous and satisfy

$$f(t) \leq a + bt^{2^*-1} \quad \forall t \geq 0 \quad \text{and} \quad \inf_{t \geq 0} \frac{f(t)}{t} = \gamma > 0,$$

for some $a, b \geq 0$. Prove that the problem

$$\begin{cases} -\Delta u = \lambda f(u) & \text{in } \Omega, \\ u > 0 & \text{in } \Omega, \\ u = 0 & \text{on } \partial \Omega, \end{cases}$$

has no solutions (e.g. in $H^1_0(\Omega)$) if $\lambda > \frac{\lambda_1}{\gamma}$. Hint: multiply by φ_1.

7. Let $f : \mathbb{R} \to \mathbb{R}$ satisfy $f(0) = 0$ and assume that there exist $p \in (2, 2^*]$ and $C > 0$ such that

$$|f(t) - f(s)| \le C|t - s|(|t| + |s|)^{p-2} \quad \forall t, s \in \mathbb{R}.$$

Set $F(t) = \int_0^t f(s) \, ds$. Prove that the functional $J : H^1(\mathbb{R}^N) \to \mathbb{R}$ defined by

$$J(u) = \frac{1}{2} \int_{\mathbb{R}^N} |\nabla u|^2 \, dx + \frac{1}{2} \int_{\mathbb{R}^N} u^2 \, dx - \int_{\mathbb{R}^N} F(u) \, dx$$

is in $C^{1,1}(H^1(\mathbb{R}^N))$.

8. Let J be the functional of the preceding exercise, but assume in addition that $p \in (2, 2^*)$ and that there exists $\mu > 2$ such that

$$f(t)t \ge \mu F(t) \quad \forall t \in \mathbb{R}.$$

Prove that if J is considered not on all of $H^1(\mathbb{R}^N)$ but on the subspace H_r of radial functions, then J satisfies the Palais–Smale condition. (*Hint*: use the compactness of the embedding of H_r in $L^p(\mathbb{R}^N)$ for $p \in (2, 2^*)$).

9. Assume that f satisfies all the assumptions of the preceding exercise, and moreover that there exists $t_0 > 0$ such that $F(t_0) > 0$. Using the result of the previous two exercises, prove that the problem

$$\begin{cases} -\Delta u + u = f(u), \\ u \in H_r \end{cases}$$

admits a nontrivial and nonnegative solution via the Mountain Pass Theorem.

10. Prove that the functional J_λ defined in (4.44) satisfies the Palais–Smale condition.

4.7 Bibliographical Notes

- Section 4.1: The Palais–Smale compactness condition was introduced in infinite dimensional Critical Point Theory in a series of works by Palais and Smale, in a slightly different form, see [40]. Our definition is the most used in the literature. For more precise, quantitative versions of the Deformation Lemma, see for example the books by Rabinowitz [43] and Struwe [45].

- Section 4.2: A most general form of the minimax principle can be found in the work of Palais [38]. The books [2, 26, 35, 43, 45, 48] all work extensively with minimax classes and should be consulted to see how the theory presented in these notes evolves. The original paper by Ambrosetti and Rabinowitz [5], already contains different examples, including "dual" minimax classes.

- Section 4.3: The Mountain Pass and Saddle Point theorems admit many variants and extensions. Particularly useful are the versions that deal with even functionals (i.e. such that $I(-u) = I(u)$). In this setting the functional is likely to possess infinitely many distinct critical points; see [43] or [45]. These kind of results are particular cases of theories dealing with functionals invariant under the action of some compact group of transformations (index theories). The interested reader can consult the quoted books [43, 45] for readable expositions of a subject that soon evolves towards algebraic topology.

- Section 4.4: The Mountain Pass Theorem applies of course to more general problems than the ones presented in these notes, for instance those where the nonlinearity also depends on x. The proofs are more technical, but up to a certain point nothing new appears. Check [43] or [8] for more examples of this kind. The functionals associated to many problems with critical growth also satisfy the geometrical assumptions of the Mountain Pass Theorem; it is the Palais–Smale condition that fails. Struwe [46] (reported in [45]) characterized all the levels at which the Palais–Smale condition is not satisfied for problem (4.17), with $p = 2^*$ (and $q(x) \equiv 0$).

  Asymptotically linear problems: the Landesman–Lazer conditions are just an example of a sufficient condition for the existence of a solution in a resonant problem; the paper [28] was the first to consider resonance. Other resonant problems that can be treated include cases where, roughly, f tends to zero at infinity and F tends to infinity (see Ahmad, Lazer and Paul [1]), or F tends to a constant ("strong resonance", see Bartolo, Benci, and Fortunato [7]), or cases where f is periodic (Solimini [44]).

- Section 4.5: Problems with parameters are studied intensively, by means of variational techniques or bifurcation methods, or a combination of the two. An introductory exposition, with many examples, can be found in [2, 4], together with an ample bibliography.

Index of the Main Assumptions

In this book we use many different hypotheses on the data of the problem currently under study. Some of them differ from each other only slightly, for technical reasons. It may happen that some assumptions apply to different problems, and in this case they are invoked many pages after they had been originally stated. For the convenience of the reader we provide here an index of all the assumptions that are used in this book so that these can be quickly looked up in this page.

(f_1) $f : \mathbb{R} \to \mathbb{R}$ is of class C^1 and is odd.

(f_2) $f'(0) = 0$ and there exists $p \in (2, 2^*)$ such that $\limsup_{t \to +\infty} \frac{|f'(t)|}{t^{p-2}} < +\infty$.

(f_3) There exists $\mu > 2$ such that $f(t)t \geq \mu F(t)$ for all $t \in \mathbb{R}$.

(f_4) For every $t \in (0, +\infty)$ there results $f'(t)t > f(t)$.

(f_5) The inequality $f'(t)t \geq (\mu - 1)f(t) > 0$ holds for every $t \in (0, +\infty)$.

(f_6) There exist $p \in (2, 2^*)$ and $C > 0$ such that for all $s, t \in \mathbb{R}$,

$$|f(t) - f(s)| \leq C|t - s|\,(|s| + |t| + 1)^{p-2}.$$

(f_7) $\lim_{t \to 0} \frac{f(t)}{t} = 0$.

(f_8) There exist $M > 0$ and $\mu > 2$ such that $f(t)t \geq \mu F(t)$ when $|t| \geq M$.

(f_9) There exists $t_0 \in \mathbb{R}$ with $|t_0| \geq M$ such that $F(t_0) > 0$.

(f_{10}) $f \in C^1(\mathbb{R})$, f is bounded and $|f'(t)| \leq C(1 + |t|^{2^*-2})$.

(g_1) $g : \mathbb{R} \to \mathbb{R}$ is of class C^1 and is odd.

(g_2) $g'(0) = 0$ and there exists $\theta \in (2, \mu)$ such that $\limsup_{t \to +\infty} \frac{|g'(t)|}{t^{\theta-2}} < +\infty$.

(g_3) The inequalities $g(t) \geq 0$ and $g'(t)t \leq (\mu - 1)g(t)$ hold for every $t \in (0, +\infty)$.

(h_1) $\Omega \subset \mathbb{R}^N$ is bounded and open, $q \in L^\infty(\Omega)$ and $q(x) \geq 0$ a.e. in Ω.

(h_2) $h \in L^2(\Omega)$.

(h_3) $f : \mathbb{R} \to \mathbb{R}$ is continuous and bounded.

(h_4) $f : \mathbb{R} \to \mathbb{R}$ is continuous and there exist $\sigma \in (0, 1)$ and $a, b > 0$ such that

$$|f(t)| \leq a + b|t|^\sigma \quad \forall t \in \mathbb{R}.$$

(h_5) $f : \mathbb{R} \to \mathbb{R}$ is continuous and there exist $a > 0$ and $b \in (0, \lambda_1)$ such that

$$|f(t)| \leq a + b|t| \quad \forall t \in \mathbb{R}.$$

M. Badiale, E. Serra, *Semilinear Elliptic Equations for Beginners*, Universitext, DOI 10.1007/978-0-85729-227-8, © Springer-Verlag London Limited 2011

(h_6) $f : \mathbb{R} \to \mathbb{R}$ is continuous and there exist $a, b > 0$ such that

$$|f(t)| \leq a + b|t|^{2^*-1} \quad \forall t \in \mathbb{R}.$$

Moreover

$$f(s)s \leq 0 \quad \forall s \in \mathbb{R}.$$

(h_7) There exist an integer $\nu \geq 1$ and $\alpha, \beta \in \mathbb{R}$ such that

$$\lambda_\nu < \alpha \leq \frac{f(s) - f(t)}{s - t} \leq \beta < \lambda_{\nu+1} \quad \forall s, t \in \mathbb{R}.$$

(q_1) $\inf_{x \in \mathbb{R}^N} q(x) > 0$ and $\lim_{|x| \to +\infty} q(x) = +\infty$.

(q_2) q is continuous and $0 < \delta := \inf_{\mathbb{R}^N} q < \sup_{\mathbb{R}^N} q =: \alpha < +\infty$.

(q_3) $\lim_{|x| \to +\infty} q(x) = \alpha$.

References

1. S. Ahmad, A.C. Lazer, J.L. Paul, Elementary critical point theory and perturbations of elliptic boundary value problems at resonance. Indiana Univ. Math. J. **25**(10), 933–944 (1976)
2. A. Ambrosetti, A. Malchiodi, *Nonlinear Analysis and Semilinear Elliptic Problems.* Cambridge Studies in Advanced Mathematics, vol. 104 (Cambridge University Press, Cambridge, 2007)
3. A. Ambrosetti, A. Malchiodi, *Perturbation Methods and Semilinear Elliptic Problems on* $\mathbb{R}^n$. Progress in Mathematics, vol. 240 (Birkhäuser, Basel, 2006)
4. A. Ambrosetti, G. Prodi, *A Primer of Nonlinear Analysis.* Cambridge Studies in Advanced Mathematics, vol. 34 (Cambridge University Press, Cambridge, 1995). Corrected reprint of the 1993 original
5. A. Ambrosetti, P.H. Rabinowitz, Dual variational methods in critical point theory and applications. J. Funct. Anal. **14**, 349–381 (1973)
6. A. Bahri, J.-M. Coron, On a nonlinear elliptic equation involving the critical Sobolev exponent: the effect of the topology of the domain. Commun. Pure Appl. Math. **41**(3), 253–294 (1988)
7. P. Bartolo, V. Benci, D. Fortunato, Abstract critical point theorems and applications to some nonlinear problems with "strong" resonance at infinity. Nonlinear Anal. **7**(9), 981–1012 (1983)
8. T. Bartsch, Z.-Q. Wang, M. Willem, The Dirichlet problem for superlinear elliptic equations, in *Stationary Partial Differential Equations.* Handbook of Differential Equations, vol. II (Elsevier/North-Holland, Amsterdam, 2005), pp. 1–55
9. H. Berestycki, P.-L. Lions, Nonlinear scalar field equations. I. Existence of a ground state. Arch. Ration. Mech. Anal. **82**(4), 313–345 (1983)
10. H. Berestycki, P.-L. Lions, Nonlinear scalar field equations. II. Existence of infinitely many solutions. Arch. Ration. Mech. Anal. **82**(4), 347–375 (1983)
11. H. Brezis, *Analyse fonctionnelle. Théorie et applications.* Collection Mathématiques Appliquées pour la Maîtrise (Masson, Paris, 1983)
12. H. Brezis, Some variational problems with lack of compactness, in *Nonlinear Functional Analysis and Its Applications, Part 1,* Berkeley, CA, 1983. Proceedings of Symposium in Pure Mathematics, Part 1, vol. 45 (American Mathematical Society, Providence, 1986), pp. 165–201
13. H. Brezis, Elliptic equations with limiting Sobolev exponents—the impact of topology. Commun. Pure Appl. Math. **39**(S1), S17–S39 (1986)
14. H. Brezis, L. Nirenberg, Positive solutions of nonlinear elliptic equations involving critical Sobolev exponents. Commun. Pure Appl. Math. **36**(4), 437–477 (1983)
15. J.-M. Coron, Topologie et cas limite des injections de Sobolev. C. R. Acad. Sci. Paris Sér. I Math. **299**(7), 209–212 (1984)

16. N.E. Dancer, A note on an equation with critical exponent. Bull. Lond. Math. Soc. **20**(6), 600–602 (1988)
17. K. Deimling, *Nonlinear Functional Analysis* (Springer, Berlin, 1985)
18. L.C. Evans, *Partial Differential Equations*. Graduate Studies in Mathematics, vol. 19 (American Mathematical Society, Providence, 1998)
19. I. Ekeland, R. Temam, *Analyse convexe et problèmes variationnels*. Collection Etudes Mathématiques (Dunod/Gauthier-Villars, Paris, 1974)
20. J.P. Garcia Azorero, I. Peral Alonso, Existence and nonuniqueness for the p-Laplacian: nonlinear eigenvalues. Commun. Partial Differ. Equ. **12**(12), 1389–1430 (1987)
21. E. Giusti, *Direct Methods in the Calculus of Variations* (World Scientific, River Edge, 2003)
22. D. Gilbarg, N.S. Trudinger, *Elliptic Partial Differential Equations of Second Order*. Classics in Mathematics (Springer, Berlin, 2001). Reprint of the 1998 edition
23. H.H. Goldstine, *A History of the Calculus of Variations from the 17th Through the 19th Century*. Studies in the History of Mathematics and Physical Sciences, vol. 5 (Springer, New York, 1980)
24. Q. Han, F. Lin, *Elliptic Partial Differential Equations*. Courant Lecture Notes in Mathematics, vol. 1 (American Mathematical Society, Providence, 1997)
25. S. Hildebrandt, A. Tromba, *The Parsimonious Universe. Shape and Form in the Natural World* (Copernicus, New York, 1996)
26. O. Kavian, *Introduction à la théorie des points critiques et applications aux problèmes elliptiques*. Mathématiques & Applications, vol. 13 (Springer, Paris, 1993)
27. I. Kuzin, S. Pohozaev, *Entire Solutions of Semilinear Elliptic Equations*. Progress in Nonlinear Differential Equations and Their Applications, vol. 33 (Birkhäuser, Basel, 1997)
28. E.M. Landesman, A.C. Lazer, Nonlinear perturbations of linear elliptic boundary value problems at resonance. J. Math. Mech. **19**, 609–623 (1969/1970)
29. E.H. Lieb, M. Loss, *Analysis*, 2nd edn. Graduate Studies in Mathematics, vol. 14 (American Mathematical Society, Providence, 2001)
30. P. Lindqvist, *Notes on the p-Laplace equation*, Report, University of Jyväskylä, Department of Mathematics and Statistics, 102 (2006)
31. P.-L. Lions, The concentration-compactness principle in the calculus of variations. The locally compact case. I. Ann. Inst. H. Poincaré Anal. Non Linéaire **1**(2), 109–145 (1984)
32. P.-L. Lions, The concentration-compactness principle in the calculus of variations. The locally compact case. II. Ann. Inst. H. Poincaré Anal. Non Linéaire **1**(4), 223–283 (1984)
33. P.-L. Lions, The concentration-compactness principle in the calculus of variations. The limit case. I. Rev. Mat. Iberoam. **1**(1), 145–201 (1985)
34. P.-L. Lions, The concentration-compactness principle in the calculus of variations. The limit case. II. Rev. Mat. Iberoam. **1**(2), 45–121 (1985)
35. J. Mawhin, M. Willem, *Critical Point Theory and Hamiltonian Systems*. Applied Mathematical Sciences, vol. 74 (Springer, New York, 1989)
36. Z. Nehari, Characteristic values associated with a class of non-linear second-order differential equations. Acta Math. **105**, 141–175 (1961)
37. W.-M. Ni, A nonlinear Dirichlet problem on the unit ball and its applications. Indiana Univ. Math. J. **31**(6), 801–807 (1982)
38. R. S Palais, The principle of symmetric criticality. Commun. Math. Phys. **69**(1), 19–30 (1979)
39. S. Pohožaev, Eigenfunctions of the equation $\Delta u + \lambda f(u) = 0$. Sov. Math. Dokl. **6**, 1408–1411 (1965)
40. R.S. Palais, S. Smale, A generalized Morse theory. Bull. Am. Math. Soc. **70**, 165–172 (1964)
41. M.H. Protter, H.F. Weinberger, *Maximum Principles in Differential Equations* (Springer, New York, 1984). Corrected reprint of the 1967 original
42. P.H. Rabinowitz, Some minimax theorems and applications to nonlinear partial differential equations, in *Nonlinear Analysis (Collection of Papers in Honor of Erich H. Rothe)* (Academic Press, New York, 1978), pp. 161–177
43. P.H. Rabinowitz, *Minimax Methods in Critical Point Theory with Applications to Differential Equations*. CBMS Regional Conference Series in Mathematics, vol. 65 (American Mathematical Society, Providence, 1986)

44. S. Solimini, On the solvability of some elliptic partial differential equations with the linear part at resonance. J. Math. Anal. Appl. **117**(1), 138–152 (1986)

45. M. Struwe, *Variational Methods. Applications to Nonlinear Partial Differential Equations and Hamiltonian Systems*, 4th edn. Ergebnisse der Mathematik und ihrer Grenzgebiete, 3. Folge [Results in Mathematics and Related Areas, 3rd Series. A Series of Modern Surveys in Mathematics], vol. 34 (Springer, Berlin, 2008)

46. M. Struwe, A global compactness result for elliptic boundary value problems involving limiting nonlinearities. Math. Z. **187**(4), 511–517 (1984)

47. S. Terracini, On positive entire solutions to a class of equations with a singular coefficient and critical exponent. Adv. Differ. Equ. **1**(2), 241–264 (1996)

48. M. Willem, *Minimax Theorems*. Progress in Nonlinear Differential Equations and Their Applications, vol. 24 (Birkhäuser, Boston, 1996)

Index

M. Badiale, E. Serra, *Semilinear Elliptic Equations for Beginners*, Universitext,
DOI 10.1007/978-0-85729-227-8, © Springer-Verlag London Limited 2011